# The Cerebellum

## Learning Movement, Language, and Social Skills

# The Cerebellum

## Learning Movement, Language, and Social Skills

**Dianne M. Broussard**

WILEY Blackwell

*Editorial offices:* 1606 Golden Aspen Drive, Suites 103 and 104, Ames, Iowa 50010, USA
The Atrium, Southern Gate, Chichester, West Sussex, PO19 8SQ, UK
9600 Garsington Road, Oxford, OX4 2DQ, UK

For details of our global editorial offices, for customer services and for information about how to apply for permission to reuse the copyright material in this book please see our website at www.wiley.com/wiley-blackwell.



*Library of Congress Cataloging-in-Publication Data*

Broussard, Dianne M., author.
The cerebellum : learning movement, language, and social skills / Dianne M. Broussard.
p. ; cm.
Includes bibliographical references and index.
ISBN 978-1-118-12563-2 (cloth : alk. paper) – ISBN 978-1-118-73007-2 (emobi) –
ISBN 978-1-118-73013-3 – ISBN 978-1-118-73025-6 (epdf) – ISBN 978-1-118-73034-8 (epub)
I. Title.
[DNLM: 1. Cerebellum. 2. Executive Function. 3. Neural Pathways–physiology.
4. Psychomotor Performance–physiology. WL 320]
QP379
612.8′27–dc23

2013027959

A catalogue record for this book is available from the British Library.

Wiley also publishes its books in a variety of electronic formats. Some content that appears in print may not be available in electronic books.

Cover images: Background art: Margaret Belknap; Top right: Dianne M. Broussard; Bottom left: © istock photo: CEFutcher; Bottom right: © istock photo: PicturePartners
Cover design by Modern Alchemy LLC

Set in 9.5/12.5pt Palatino by Aptara® Inc., New Delhi, India

1 2014

# Contents

# Acknowledgments

I would like to thank my teachers for having such different views about the same part of the brain. Maurizio Mirolli introduced me to the cerebellum. Hiroharu Noda, Bob McCrea, and Steve Lisberger made it clear that the cerebellum is a worthwhile and fascinating problem. I would also like to thank Helen Bronte-Stewart for her beautifully clear explanation of Marr's hypothesis, long before I found the time to actually read his paper; and Tom Masino, for asking me, "So what does the cerebellum do, anyway?" It took a bit of time, but here is my answer.

All of my colleagues and students have inspired me to write this book; it is really for them, and for their students. I thank Chris de Zeeuw, Jennifer Raymond, Kathy Cullen, Jim Sharpe, Dave Tomlinson, Doug Tweed, Sascha du Lac, Dora Angelaki, Albert Fuchs, Rich Krauzlis, Fred Miles, Jim McElligott, and many others, for asking good questions. I especially thank Jerry Simpson, for being skeptical about the cognitive function of the cerebellum.

My husband and colleague, David R. Hampson, first made me aware of the need for this book by getting me involved in his research on autism spectrum disorders. He has helped in innumerable ways. Thank you especially for the many hours of reading, the feedback, and the many discussions.

Finally, I am tremendously grateful to my daughter, Luci Belknap, for her constant support, encouragement, and sympathy through the long process of writing, and for the cover art.

# Introduction

Early modern humans had a problem with brain size. These Stone Age humans probably had very high maternal and infant mortality, even higher than in the so-called primitive societies of modern humans, thanks to their larger crania. The expansion of the cerebral hemispheres had also made it necessary for babies to be born at a more immature stage than modern humans are. What if an increase in the size of the cerebellum, which was relatively small in Cro-Magnon Man, could improve the efficiency of the human brain, allowing the cerebral hemispheres and the diameter of the cranium to become slightly smaller, while maintaining the competitive edge provided by human intelligence? Although we do not know if this in fact did happen, it is consistent with what we do know (Weaver 2005). Such an improvement could have allowed more infants, and mothers, to survive childbirth while also allowing infants to be more mature at birth.

The cortex of the cerebellum is a huge, multilayered sheet of neurons that is folded like an accordion. The folds are compressed into a structure resembling a "little brain," which lies behind and beneath the cerebral hemispheres. But it is not really so little. In humans, if all of its folds were flattened out, the cerebellar cortex would extend for more than 1 m from front to back (Braitenberg & Atwood 1958). Several million nerve fibers exit the cerebellum (Glickstein *et al.* 2011). What is the function of all of this processing power and connectivity? What does the cerebellum do? In this book, I will argue that the cerebellum is a supplementary processing device that boosts the computing power of the cerebral cortex—and that it can be used for essentially any task.

It has been said that people born without a cerebellum are nearly normal, but this is a myth. In fact, the few patients with "cerebellar agenesis" have symptoms resembling severe cerebral palsy. In all known cases, their deficits include severe motor disability and profound mental retardation. What is more, the cerebellum is not completely lacking in any of them; some part of it always remains (Glickstein 1994). In fact, the number of cases where the cerebellum has been confirmed to be completely lacking in an individual who survived infancy is zero.

Individuals can survive without most of their cerebellum, but they need a lot of help. Also, we can walk and talk without parts of the cerebellum, just not very well. The cerebral cortex is plastic, and can learn without the cerebellum, and even (to some extent) can compensate for its absence. But having a cerebellum allows us to speed up, perfect, and extend our behavioral repertoire. For animals in the wild (and even occasionally for modern humans), speed is absolutely crucial for survival. Good motor performance can be a matter of life and death.

The first goal of this book is to give a general overview of cerebellar function: what it does, and how it does it. Section I will focus on how the cerebellum works. Section II will show how the cerebellum participates in motor learning. Section III will describe the contribution of the cerebellum to precision, timing, and coordination of movement. Motor control is one of the most complicated things that animals—including humans—do, and the motor functions of the cerebellum allow us to interact promptly and successfully with our environment.

But the cerebellum also has other functions that have nothing to do with motor control. As we will see in Section IV, these include certain aspects of cognition: language, working memory, and attention as well as certain emotional and social functions. More cerebellar functions almost certainly remain to be discovered. There have been difficulties obtaining evidence for nonmotor cerebellar functions, mostly because we are talking about faculties that are exclusively human. The quality of the evidence is improving rapidly, but many clinicians and neuroscientists still believe that "the cerebellum is for motor control." My second goal is to demonstrate that this view should be changed.

## REFERENCES

Braitenberg, V. & Atwood, R.P. (1958) Morphological observations on the cerebellar cortex. *J. Comp. Neurol.*, **109**, 1–33.

Glickstein, M. (1994) Cerebellar agenesis. *Brain*, **117**, 1209–1212.

Glickstein, M., Sultan, F. & Voogd, J. (2011) Functional localization in the cerebellum. *Cortex*, **47**, 59–80.

Weaver, S.H. (2005) Reciprocal evolution of the cerebellum and neocortex in fossil humans. *Proc. Natl. Acad. Sci. USA*, **102**, 3576–3580.

# Section I

## The Neuronal Machine

*The Cerebellum: Learning Movement, Language, and Social Skills*, First Edition. Dianne M. Broussard.

# 1 Structure and Physiology

## ANATOMY OF THE CEREBELLAR CORTEX

Even the most primitive vertebrates have a cerebellum. For example, the cyclostomes (hagfish and lampreys) have a cerebellum, even though—like other fish—they do not have a cerebral cortex.

The cerebellum may have first appeared as a computational device for the lateral line systems of fish.Lateral lines are rows of tiny hair cells on the skin of fish that detect rocks, fish, and other solid objects. Without their lateral line organs, fish collide with obstacles (Sarnat & Netsky 1974). While swimming alongside a wall, for example, the fish's movement is continually adjusted to maintain a constant distance, based on signals from the lateral lines. Both the lateral line nerves and the central nuclei associated with them send axonal projections into the cerebellum. The purpose of the first cerebellum may have been to carry out computations that allowed fish to use the sensory feedback from their lateral lines to guide swimming.

The cerebellum of cyclostomes works with a very simple structure. It contains only two types of neurons: the tiny and very numerous granule cells, and the large Purkinje cells (P-cells), with their extensive dendritic arbors. The dendrites of each P-cell branch within a flattened, nearly planar field in the molecular layer. Granule cells terminate on and excite the P-cells. P-cells are the only neurons whose axons leave the cortex. Unlike most other large projection neurons of the brain, they inhibit their target neurons.

Both granule cells and P-cells receive afferent input. Granule cells are innervated by the mossy fibers, so called because their axon terminals resemble miniature branches and leaves of moss. In cyclostomes, mossy fibers originate from the lateral line and vestibular nuclei. The P-cells have direct input from the ivy-like climbing fibers, whose cell bodies are located in the inferior olivary nuclei of the brainstem.

Throughout vertebrate evolution, the cerebellar cortex has kept these primitive features and added more. In humans, the cerebellar cortex has three layers (Figure 1.1): the molecular layer, which is a surface layer

*The Cerebellum: Learning Movement, Language, and Social Skills*, First Edition. Dianne M. Broussard.

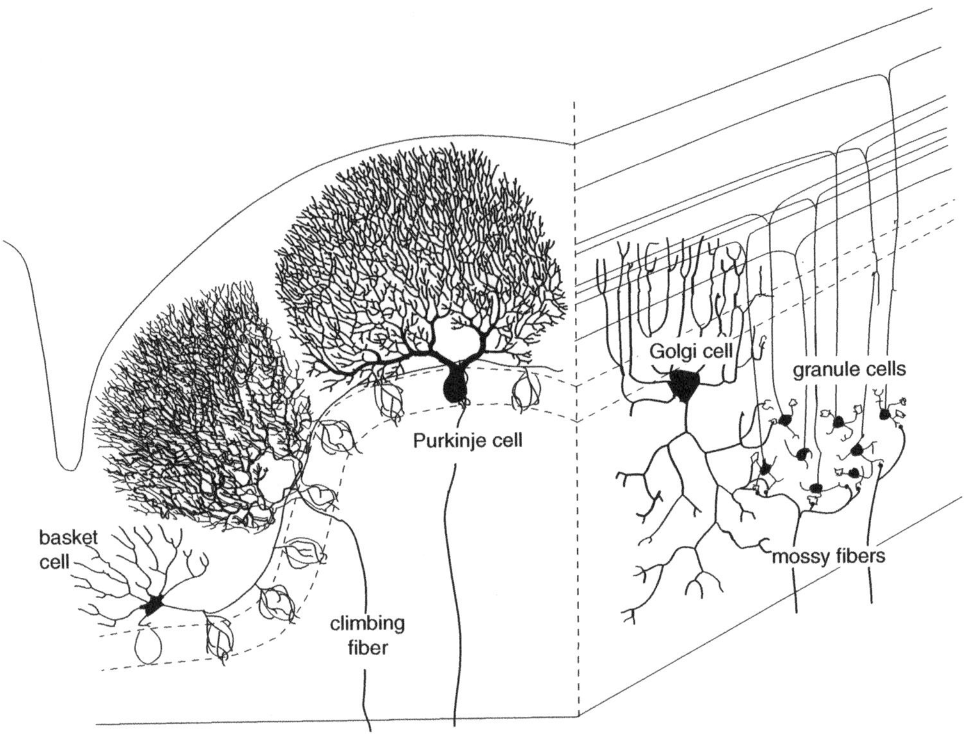


**Figure 1.1** Human cerebellar cortex. A P-cell, a climbing fiber, and a basket cell are shown in the sagittal plane. Golgi and granule cells and mossy fibers are shown in the coronal plane. The dashed lines indicate the P-cell layer. The axon of the basket cell is shown extending in the sagittal plane and surrounding the cell bodies of several P-cells. See text for other details.

containing mostly axons; the P-cell layer; and the granular layer. The granular layer contains between $10^{10}$ and $10^{11}$ granule cells in humans (Braitenberg & Atwood 1958).[1]

The axons of granule cells extend from the granule cell layer, through the P-cell layer, and into the molecular layer where they bifurcate, each branch making a right-angle turn. The branches, which are thin and unmyelinated, are called parallel fibers. They extend for several millimeters, terminating extensively on the P-cell dendrites and releasing glutamate. Input from the mossy fibers can excite P-cells through the granule cells and parallel fibers. The rectangular lattice formed by the P-cell dendritic arbors and the parallel fibers suggests an efficient device for processing and/or storing information.

[1]This astounding number represents most of the neurons in the brain. Although it has been questioned, these authors were very careful in their methodology.

The climbing fibers originate from the inferior olive. In general, each P-cell receives input from only one climbing fiber, although there are exceptions (Nishiyama & Linden 2004). Climbing fibers usually extend through the entire thickness of the molecular layer, twining around the dendrites of the P-cell.

In mammals, the cerebellar cortex contains inhibitory interneurons, that form another link between the granule cells and the P-cells. These are the basket cells and the stellate cells. The basket cell is named for the structure of its terminal, which is a giant synapse, also known as the "pinceau". One pinceau surrounds the soma of each P-cell, secreting GABA and inhibiting the initial segment of the axon as well as the soma itself. The long axons of the basket cells extend parallel to the P-cell dendritic arbors, at right angles to the parallel fibers. The discharge pattern of a basket cell effectively sculpts the discharge patterns of each P-cell that it innervates. Stellate cells also inhibit P-cells, but they terminate on the dendrites and are less effective.

Basket cells and stellate cells also receive collaterals from the climbing fibers. This means that each excitatory input to the P-cell is accompanied by an inhibitory input that is derived from it (Figure 1.2).

Only one type of inhibitory interneuron, the very large Golgi cell, terminates on granule cells. The Golgi cell closes an inhibitory feedback loop. As a result, like P-cells, granule cells receive both excitatory and inhibitory inputs.

In addition to mossy fibers, which are glutamatergic and terminate exclusively in the granular layer, and climbing fibers (also glutamatergic), input pathways to the cerebellum arise from the locus coeruleus (adrenergic neurons) and raphe nuclei (serotonergic neurons). These axons terminate directly on P-cells.

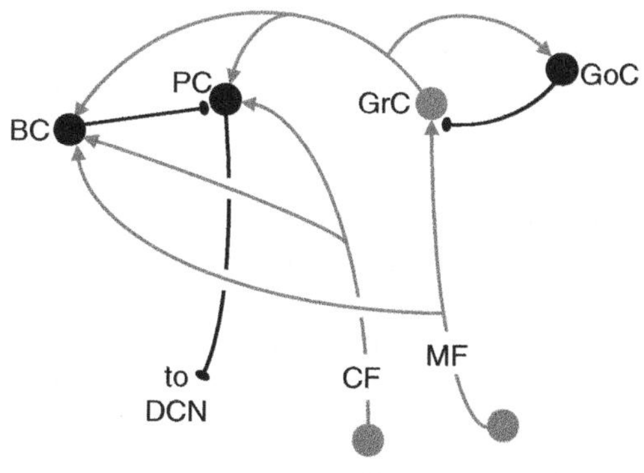


**Figure 1.2** A simplified diagram of the connections among neurons in the cerebellar cortex. Black cells and round terminals are inhibitory. Gray cells and arrows are excitatory. PC, Purkinje cell; BC, basket or stellate cell; GoC, Golgi cell; GrC, granule cell; MF, mossy fiber; CF, climbing fiber; DCN, deep cerebellar nucleus.

## PHYSIOLOGY OF THE CEREBELLAR CORTEX

What do all of these parts do? Over the years, neurophysiologists have carried out hundreds of studies of the cerebellar cortical circuit, with the goal of answering this question. At least one important principle has held up over time: the close relationship between the vine-like climbing fiber and the P-cell is of primary importance to cerebellar function. Because the climbing fiber has a large number of glutamatergic terminals on the P-cell, each action potential in the climbing fiber strongly depolarizes the dendritic arbor. This strong synaptic activation, along with a "resurgent" sodium current (Raman & Bean 1997), results in an action potential in the P-cell that has not one but several peaks (the "complex spike"). On the cellular level, complex spikes are necessary for cerebellar learning. In at least some cases, climbing fibers bring information about errors into the cortex.

In adult mammals, most P-cells are innervated by only one climbing fiber, and in fact, this pattern of innervation is required for certain kinds of learning (Kimpo & Raymond 2007). But climbing fibers fire infrequently, usually evoking 10 or fewer complex spikes per second. Meanwhile, the P-cell fires "simple spikes" (ordinary action potentials) steadily at up to 100 spikes per second or more, allowing it to encode rapid sensory and motor events. The steady stream of simple spikes is the main output of the cerebellar cortex, but as we shall see, complex spikes can affect the stream.

### *Parallel fibers: many weak inputs*

The sheer number of synapses that connect granule cell axons, especially the long parallel fibers, with P-cells suggests that the mossy-fiber input pathway must be important. There are roughly 150,000 synapses from parallel fibers on each P-cell, and we each have over 10 million P-cells. This arrangement provides enormous computational power. It has the potential to encode a tremendous number of different components of motor patterns, for example.

But despite these impressive numbers, it is clear that the parallel fiber–P-cell (PF–PC) circuit is not the only contributor to cerebellar signal processing in mammals, and it may not even be the most important contributor. For one thing, P-cells can generate simple spikes at a steady rate on their own, without any synaptic input (Hounsgaard & Yamamoto 1979). This "spontaneous" firing likely occurs because the resurgent sodium current does not completely inactivate. As a result, the resting potential of the P-cell is above the threshold for firing (Raman & Bean 1997). Spontaneous firing can be modulated by the many synaptic inputs to the P-cell, but inhibitory inputs are likely to be more useful for this kind of modulation than excitatory ones.[2]

## *Climbing fibers and inhibitory interneurons*

Another important contributor to P-cell discharge patterns is the climbing-fiber input. At the same time that it causes complex spikes, the climbing-fiber input actually *decreases* the rate of simple-spike firing by the same P-cell (Montarolo *et al.* 1982). In most behavioral paradigms, complex spikes and simple spikes show opposing responses, with complex spikes decreasing while simple spikes increase their discharge rates, and vice versa. In fact, the pattern of complex-spikes seems to *determine* the simple-spike responses of some P-cells to sensory stimuli (Barmack & Yakhnitsa 2003). Although we do not know exactly how this works, evidence indicates that at least two mechanisms may contribute: the activation of calcium-activated potassium channels in P-cells (McKay *et al.* 2007), and the activation of stellate and basket cells (Barmack & Yakhnitsa 2011). In at least some cerebellar regions, the stellate and basket cells control whether P-cells will increase or decrease their simple-spike firing rates during particular sensory stimuli (Barmack & Yakhnitsa 2008).

Both granule cells and P-cells do more than just add up their synaptic inputs. P-cells sometimes end up responding to sensory and motor events in a similar way to the mossy fibers, but sometimes they do not. In fact, the responses of P-cells and mossy fibers can end up being opposites (Barmack & Yakhnitsa 2008). This situation is not as strange as it first seems. Each parallel fiber provides one or a few relatively minor inputs to the P-cell. These inputs arrive on the dendritic spines, which are electrically quite distant from the spike generator of the P-cell. Furthermore, the P-cell is spontaneously active, so after each spike it may reach threshold again within a few milliseconds, whether or not it receives any depolarizing synaptic input. On the other hand, inhibitory inputs can sculpt the spontaneous activity of the P-cell. The basket cell actually terminates on the soma and the proximal axon, and override the more distal parallel-fiber inputs. It is clear that the basket cell is a key player in determining how P-cells will respond to sensory and motor events.

---

[2]This is because inhibitory inputs open (usually) chloride channels, which shunts postsynaptic excitatory currents and also intrinsically-generated currents. This is roughly analogous to a short in electrical wiring and has a similarly powerful effect.

While the climbing fibers and inhibitory interneurons shape P-cell discharge patterns, the granule-cell input to P-cells may be doing something else entirely. For example, it could function by toggling the activity levels of P-cells between up and down states (in other words, states of high activity and silence) (Rokni *et al.* 2009). Because in the past, experiments have been based on the assumption that the parallel fibers determine the P-cell discharge pattern, researchers now need to take a step back and find out exactly what their function is and how it is implemented.

## *Patterns of synaptic input*

Although parallel-fiber inputs apparently do not dominate P-cell firing, they still may contribute in unique ways. For example, parallel fibers can distribute information laterally, at right angles to the basket cell axons. This arrangement could, for example, add to behavioral flexibility by permitting learning of a variety of input–output transformations within the same storage space.

In the early 1900s, Santiago Ramón y Cajal noticed that the parallel fibers and the long axons of the inhibitory basket cells are oriented at right angles to each other, with the parallel fibers extending mediolaterally and the basket-cell axons extending rostrocaudally (Figure 1.1). This observation led to the idea, originally proposed by Janos Szentagothai (Eccles *et al.* 1967), that excitation of a "beam" of P-cells aligned with a group of parallel fibers, combined with lateral inhibition adjacent to the beam, is an important mechanism of cerebellar cortical processing. This idea has been intensely debated in recent years. Clearly, the beam of activation is present during stimulation of the cerebellar surface and is flanked by inhibition (Cohen & Yarom 2000; Gao *et al.* 2006; Wang *et al.* 2011). But to make a long story mercilessly short, the beam is probably a laboratory artifact. The pattern *in vivo* consists of activated patches of cortex, rather than beams (Brown & Ariel 2009). This could occur because P-cell excitation is dominated by the terminals given off by the ascending segments of axons from a cluster of granule cells (Cohen & Yarom 1998).

A puzzling, but clearly important feature of the cortical circuit is the presence of the Golgi cells. These large inhibitory neurons receive input from the parallel fibers and the climbing fibers, and inhibit granule cells. The axons of Golgi cells are restricted to a parasagittal zone; that is, they run more or less parallel to the basket-fiber axons, and at right angles to the parallel fibers (Barmack & Yakhnitsa 2008).

The most interesting thing about the Golgi cells is the way that they terminate. Axon terminals of Golgi cells and mossy fibers, along with the claw-like dendrites of granule cells, combine to form "glomeruli" (a more beautiful word for clumps). The glomeruli act as containers to retain neurotransmitters after their release from two kinds of terminals. This seems to change the temporal properties of neurotransmission (Mitchell & Silver 2000; DiGregorio & Nusser 2002). The effect of this whole arrangement is that the inhibitory pathway (via the Golgi cells) can regulate excitatory transmission between mossy fibers and granule cells directly, regulating parallel-fiber signals right at their source (Figure 1.2).

Golgi neurons are spontaneously active, and recently, it was found that they can be silenced by peripheral sensory input (possibly by way of the stellate and/or basket cells). This suggested that Golgi cells may act as "associative filters," detecting particular situations in which it is desirable to allow sensory input to flow into the cortex by way of the mossy-fiber pathway (Holtzman *et al.* 2006). For example, in the eye-movement system, Golgi cells have been proposed to contribute to high-pass filtering, a process that rejects unchanging or slowly changing components of the sensory input (Miles *et al.* 1980). Like the output stage, the input stage of the cerebellar cortex is under inhibitory control.

Finally, in addition to providing the output of the cerebellar cortex, the P-cells contribute to the function of the cortical circuit. Some of the dynamic properties of the circuit rely on intrinsic characteristics of the P-cells, in particular their membrane potassium channels (Engbers *et al.* 2012).

## *Firing in synchrony*

Some neurons in the cerebellum and inferior olive fire together as synchronous groups, under the right conditions. The gap-junction protein, connexin,[3] is present in most neurons of the inferior olive, the deep cerebellar nuclei (DCN), and also the inhibitory interneurons of the cerebellar cortex (van der Giessen *et al.* 2006), suggesting that these neuronal types form continuous networks that fire together. The inferior olive, in particular, probably has a higher density of gap junctions than any other brain region. Synchronous firing has great potential for signal processing, as we will see in subsequent chapters.

In addition to their targets outside the cerebellar cortex, P-cells project recurrently to their neighbors. This recurrent inhibition appears to synchronize P-cell simple-spike firing across the population (de Solages *et al.* 2008). As a result, although P-cells do not express connexins, synchronous firing of P-cells may still have a role in accurate timing of movements.

---

[3]Connexin is not a single protein but several proteins that are distinguished by numbers. Connexin45 is present throughout the cerebellum, but it may be connexin36 that actually is crucial for the formation of gap junctions. See the paper by van der Giessen *et al.* (2006) for details.

From feedforward inhibition by the basket cells, to feedback inhibition by Golgi cells at the input stage, to feedback by P-cells at the last possible moment, inhibition is the key to cerebellar cortical function. This principle also holds in the brain regions that receive input from P-cells. But before looking at those, we need to get a better idea of the anatomy of the cerebellum as a whole.

## SUBDIVISIONS OF THE CEREBELLUM

Unlike the cerebral cortex, the cerebellum retains the very stereotyped cytoarchitecture shown in Figure 1.1 in all of its subdivisions. However, the different cerebellar regions have different functions, mediated by their different input and output connections.

There are two major ways to subdivide the cerebellum. First, we can divide the cerebellum up along crosswise lines, into lobes. Second, we can divide it lengthwise, into zones. Both of these schemes are meaningful in terms of cerebellar function.

First, the lobes. In mammals, there are three: the vestibulocerebellum, the anterior lobe, and the posterior lobe (Figure 1.3). The lobes are separated by deep fissures which run across the cerebellum from side to side. The lobes are diagrammed in Figure 1.3b. Each lobe is subdivided into many folia, giving the cerebellum its characteristic wrinkled appearance (Figure 1.3a).

The vestibulocerebellum is the oldest of the three lobes. It is closely linked to vestibular function and is responsible for the learning capacity of our vestibular reflexes. The cortex of the vestibulocerebellum is interconnected directly with the vestibular nuclei. The four vestibular nuclei are located in the brainstem and receive fibers from the vestibular division of the eighth cranial nerve (auditory nerve). The vestibular division carries information from the vestibular labyrinth, related to the position and motion of the head.

In some fish and amphibians, the spinal cord is also interconnected directly with the cerebellar cortex, specifically with the corpus cerebelli. Other vertebrates, including humans, do not have a corpus cerebelli but we do have an anterior lobe, which has indirect connections with the spinal cord.

Possibly the most interesting region of the cerebellum, the posterior lobe, evolved relatively recently. Only mammals have a posterior lobe, and it is greatly expanded in humans (Figure 1.3a). The posterior lobe is linked with many regions of cerebral cortex, including association cortex, by way of the pontine nuclei of the brainstem. Recently, advanced imaging methods have been used to investigate the human posterior lobe. This approach has revealed many possible functions, such as language and other cognitive skills, including social skills.

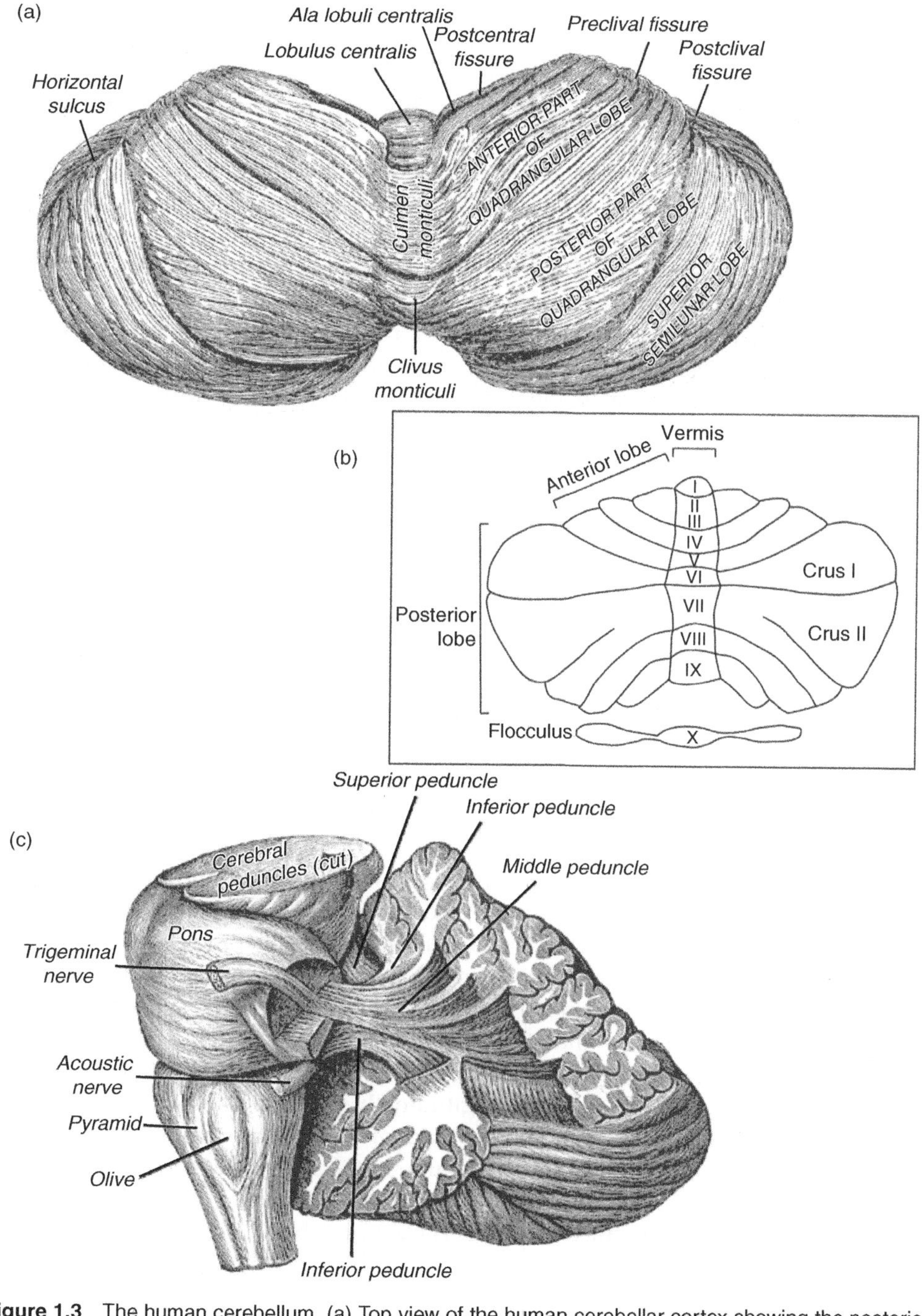


**Figure 1.3** The human cerebellum. (a) Top view of the human cerebellar cortex showing the posterior lobe (Source: Gray's Anatomy, Fig. 702). (b) A simplified map of the cerebellar cortex. The Roman numerals are the numbers of lobules described in the text. Most of the Latin names have been omitted. Lobule X is the vestibulocerebellum, and is shown separately. (c) The relationship between the brainstem and the cerebellar cortex, in cutaway view, showing some of the cerebellar lobules. The peduncles are fiber tracts connecting the cerebellum to the rest of the brain (Source: Gray's Anatomy, Fig. 705).

The three major subdivisions of the cerebellum have been defined differently by some authors. Their divisions are called the archicerebellum, paleocerebellum, and neocerebellum.[4]

Either way, the lobes are further subdivided into lobules or folia ("leaves"), which are numbered I–X (I–V, anterior lobe; VI–VII and part of IX, posterior lobe; X and part of IX, vestibulocerebellum) (Figure 1.3). This system is convenient for finding one's way around the cerebellum, and is consistent across different species of mammals (including humans). However, even within a lobule, the more medial and more lateral regions may have very different functions and connections, leading to some rather esoteric Latin terminology which is *not* the same across species. The interested reader can find detailed maps of the cerebellum, with all of its Latin labels, in several textbooks, such as Brodal (1981) or Carpenter (1976). For the most part, in this book, I will use the simpler numerical system rather than the Latin names when referring to lobules.

The second way to map the cerebellar cortex is by dividing it into central, intermediate, and lateral regions. The central region, which extends down the middle of the cerebellum from front to back (including all of the lobes), looks like a segmented worm and is called, accordingly, the vermis. Lateral to the vermis is the paravermis, and the most lateral zone is the cerebellar hemisphere, which is by far the largest cerebellar zone in humans. Based on their anatomical connections, these regions can be subdivided into roughly 12 longitudinal, or parasagittal zones. These zones are narrow, but some of them extend through all folia of the anterior and posterior lobes, as illustrated in Apps & Hawkes (2009). Each parasagittal zone of the cortex projects to a different region of the deep cerebellar nuclei (DCN) (Voogd *et al.* 2003).

Parasagittal zones are usually defined by their output, but the same zones can also be distinguished based on the climbing-fiber input connections and even, to some extent, on mossy-fiber inputs. Even the axons of stellate and basket cells and the dendrites of the larger neurons in the cerebellar cortex extend only within individual zones, suggesting that the zones operate as independent modules (De Zeeuw *et al.* 2011).

Detailed analysis of the cerebellar zones has shown that they can be further subdivided into longitudinally organized microzones that are on the order of 0.1 mm wide (Oscarsson 1979). Microzones may be the real functional units of the cerebellum. In the same size range, "stripes" have been described that occur in pairs, one expressing and one not expressing a protein called zebrin II. The zebrin-labeled stripes exhibit enhanced

---

[4]Some authors include the vermis of the posterior lobe in the paleocerebellum, and lobule IX, which is part of the vestibulocerebellum, is excluded from the archicerebellum. For a detailed description of the cerebellar lobes, see (Brodal 1981).

basket-cell inhibition (Gao *et al.* 2006) and enhanced synaptic plasticity (Wang *et al.* 2011). In some regions of the cerebellar cortex, the pairs of stripes truly are functional units (Graham & Wylie 2012). Zebrin stripes are not present in all areas of the cerebellum, suggesting that other marker proteins may also be useful for delineating the microzones.

## THE GATEKEEPERS: VESTIBULAR AND DEEP CEREBELLAR NUCLEI

The P-cells of the cerebellar cortex have axons that leave the cortex and terminate in the vestibular nuclei and in special clusters of cells, the DCN, which are located at the base of the cerebellum. The axons of neurons in the DCN in turn send the results of cerebellar computations to the rest of the brain (Figure 1.4). The vestibular nuclei and DCN provide the only routes by which information from the cerebellum can reach the rest of the brain.

There are roughly 26 P-cells for every DCN neuron (Palkovits *et al.* 1977), and P-cells innervate multiple DCN neurons. This means that there is tremendous convergence of P-cell axon terminals, which are GABAergic, on individual neurons of the DCN. DCN neurons are spontaneously active, and their discharge patterns are shaped by their inhibitory input from the P-cells.

There are four DCN. Each one has two names, thanks to early anatomists who assigned them different names in primates and nonprimates. The medial nucleus (fastigial nucleus in primates) receives output from the vermis. The anterior interpositus (emboliform in primates) and

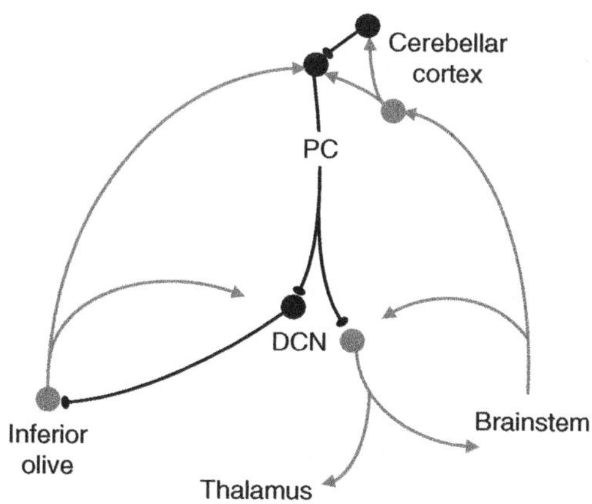


**Figure 1.4** A simplified diagram of the cerebellar connections. Gray neurons and arrows are excitatory; black neurons and elliptical terminals are inhibitory. PC: P-cell. See text for details.

posterior interpositus (globose) nuclei receive the output of the paravermis. Finally, the hemispheres project to the lateral (dentate) nuclei. The dentate nuclei are greatly enlarged in humans.

The DCN and the vestibular nuclei receive collaterals of both major groups of inputs to the cerebellum (the climbing fibers and the mossy fibers). Because of this arrangement, for each brain pathway that is linked to the cerebellum, the cortex is essentially a "side loop". In other words, one pathway proceeds directly through either the DCN or the vestibular nuclei, and an alternative pathway includes the granule cells and P-cells of the cerebellar cortex (see Figure 1.4). The side loop inhibits the DCN neurons, while the direct pathway is excitatory. By inhibiting the direct pathway, the cerebellar cortex makes precise adjustments to motor commands and other signals.

The DCN contain several types of cells, both excitatory (glutamatergic) and inhibitory (GABAergic and glycinergic). The GABAergic neurons are small, and probably include local interneurons, as well as projection neurons that reach the inferior olive. Notably, the inhibitory projection from the DCN to the inferior olive is part of a closed loop that includes the inferior olive, P-cell, and DCN neuron. Two of the cells in this loop are inhibitory, and the third is the climbing fiber, which we already know reduces the P-cell simple-spike discharge.

What this means is that by way of the inferior olive, the inhibitory DCN neurons provide feedback to the cerebellar cortex. Feedback from the loop has an important role in determining the pattern of P-cell discharge (Bengtsson *et al.* 2004). Meanwhile the glutamatergic and glycinergic neurons in the DCN, which are relatively large, project to other regions of the brain.

The P-cell terminals in the DCN are large and GABAergic, and there are a very large number of them. Why are the DCN neurons not completely overwhelmed by all of their inhibitory input? First, like P-cells they contain a special type of noninactivating sodium channel (Raman *et al.* 2000), which confers spontaneous firing and probably also helps to counteract the inhibition. Also, the inhibitory synapses from P-cells onto the DCN neurons are chronically depressed, operating at only about 10% of their capacity (Telgkamp & Raman 2002).

One of the best-established cerebellar functions is to time movements accurately. Synchronous inhibitory inputs from P-cells to DCN neurons can control the timing of DCN action potentials very precisely (Gauck & Jaeger 2000), and this mechanism may contribute in turn to the timing of movements and other brain functions. Signaling by timing DCN action potentials provides another example of the importance of inhibition in cerebellar function.

The connections of the cerebellum with the rest of the brain tell us a great deal about its function. The rest of this chapter will summarize these connections briefly.

## AFFERENT CONNECTIONS OF THE CEREBELLUM

In keeping with the early appearance of the cerebellum during vertebrate evolution—before any forebrain[5] structures appeared—all of its input signals are routed through the brainstem. The inferior olivary nucleus is the only source of climbing fibers, but mossy fibers arise from many brainstem regions, including the pontine nuclei, the reticular formation, and the vestibular nuclei. The input from the pons is by far the largest in humans[6] (the middle cerebellar peduncle; see Figure 1.3c). The pons receives information from the cerebral cortex and sends a massive projection to the cerebellum. It does not give off collaterals to other regions of the brain (Jones & Wise 1978). But this pathway does involve some processing, and there is a great deal of convergence and divergence in the pontine nuclei.

Because of its input from the pons, the cerebellum can participate in many of the major functional brain networks that have been identified using functional brain scanning in humans.

### *The inferior olive*

The inferior olive is a large, folded sheet of cells at the base of the brainstem. Probably, every cell in the olive provides one or more climbing fibers to the cerebellar cortex. As a population, climbing fibers carry several kinds of information, supplying visual, vestibular, auditory, and somatosensory signals to the P-cells and (via collaterals) to the DCN. The inferior olive also provides efference copy, that is, copies of signals that also are sent to motoneurons; these arise from the motor cortex, red nucleus, and several eye-movement centers (the superior colliculus, accessory oculomotor nuclei, and nucleus prepositus hypoglossi).

The climbing-fiber projection is very orderly. The arrangement of input to the mediolateral zones of the cerebellar cortex is a clear point-to-point projection from dedicated regions of the inferior olive to exclusive zones of the cortex (Groenewegen *et al.* 1979). The precision of the olivocerebellar projection probably allows particular regions of the olive to regulate learning of different motor (or other) functions, allowing each zone of the cerebellum to act independently in its own reverberating loop (De Zeeuw *et al.* 2011). Also, the climbing-fiber projections to the anterior and posterior parts of each zone are for the most part independent (Apps 2000).

---

[5]The forebrain includes the cerebral cortex, basal ganglia, thalamus, hypothalamus, and associated structures.

[6]Pons means "bridge". The pons resembles a bridge because of the enormous number of axons that cross the midline and ascend to the cerebellar cortex on the opposite side.

This is consistent with different functions for (for example) the vestibulocerebellum and posterior lobe.

The inferior olive projects to the entire cerebellum, but mossy fibers arise from different regions to innervate each of the three cerebellar lobes. Next, I will briefly summarize the mossy-fiber inputs to each lobe.

## *Afferents to the vestibulocerebellum*

The oldest part of the cerebellum, the vestibulocerebellum (lobules IX and X) receives mossy-fiber input directly from the vestibular division of the VIIIth cranial nerve as well as from the vestibular nuclei, paramedian reticular nuclei, and other regions of the medulla that are related to eye movements, such as the nucleus prepositus hypoglossi. The lateral part of lobule X (usually called the flocculus) also receives input from the pontine nuclei, including signals that originate from the cerebral cortex (Langer *et al.* 1985). The vestibulocerebellum is important for learning accurate eye movements and for gaze stabilization, which is necessary for clear vision while moving. It also has a nonmotor role, in the perception of self-motion.

In primates, the caudal and lateral-most part of lobule IX (the "ventral paraflocculus") is functionally part of the vestibulocerebellum, even though most diagrams of the cerebellum include it in the posterior lobe. The ventral paraflocculus contains unipolar brush cells (Mugnaini & Floris 1994), a type of neuron that is present only in the vestibulocerebellum. Unipolar brush cells form a short excitatory connection between mossy fibers and granule cells.[7]

The ventral paraflocculus receives no direct input from the vestibular nerve (only indirect input), but its other afferents are like those of the lateral vestibulocerebellum (Nagao *et al.* 2002). Like the flocculus it contains parasagittal zones, and they are continuous with those of the flocculus (Voogd & Glickstein 1998). As this implies, the flocculus and ventral paraflocculus work together in primates (Rambold *et al.* 2002). Interestingly, this is not the case in other mammals (Osanai *et al.* 1999). It is possible that because primates are highly visual and need to stabilize their gaze more precisely, the oculomotor system has recruited adjacent cerebellar areas to help with gaze stabilization.

## *Afferents to the anterior lobe*

The anterior lobe receives a large amount of sensory input directly from the dorsal horn of the spinal cord, transmitting touch, pain, pressure, and

[7]Unipolar brush cells may serve to amplify some input signals, but otherwise their function is unknown. For more information, see Russo *et al.* (2008).

especially proprioceptive or "kinesthetic"[8] signals. These primary sensory signals come from all four limbs, the trunk, and the neck, and are somatotopically organized (Grant 1962). The anterior lobe also receives vestibular information from the brainstem, which may be useful for balance and locomotion.

In addition to these primary sensory signals, the anterior lobe receives large projections from the lateral reticular nucleus of the medulla and from the red nucleus that may contain more processed sensory and/or motor signals. The red nucleus (an important premotor nucleus) also projects to the fastigial nucleus, and this projection likely represents collaterals of mossy fibers that are directed to the anterior lobe. Finally, there are projections to the anterior lobe from the pontine nuclei, relaying information from the cerebral cortex.

The anterior lobe has a known function in locomotion. In human patients, damage to the anterior lobe causes ataxia, or uncoordinated walking. In humans, walking requires a long period of practice, and the cerebellum is likely to play a part in that process.

### *Afferents to the posterior lobe*

As already mentioned, the posterior lobe is greatly expanded in humans (Figure 1.3). It has numerous functions that we know of so far, and almost certainly has more that remain to be discovered. The posterior lobe has a role in motor control, the transformation of sensory information into motor commands, and in motor learning. It also participates in movement preparation, planning, and coordination; in perception of self-motion; and in timing. Finally, in humans and other primates the posterior lobe has significant cognitive and emotional tasks.

There is abundant sensory input to the posterior lobe. It receives some direct vestibular and somatosensory afferents, but most of the input comes from the pontine nuclei and the pontine reticular formation. The pons sends several kinds of processed sensory and motor information to the posterior lobe. Motor areas of the cerebral cortex are the source of most of the input to the pons, with visual areas, temporal lobe, and parietal and frontal lobes also contributing (Brodal 1978).

The posterior-lobe vermis (lobules VI and VII) is one cerebellar region that integrates different types of information, and has motor as well as nonmotor functions. It receives, by way of the pons, information about

[8]Kinesthesia usually refers to the sense of body motion but has no exact physiological definition. Proprioception is the sense of body position and motion that is derived from the stretch receptors and Golgi tendon organs. It does *not* include vestibular sensory information; this is a common error of historical origin.

visual targets that are being tracked by the eyes. Inputs to lobules VI and VII also arise from auditory cortex. The reticular tegmental nucleus of the pons ("NRTP")[9] also projects heavily to lobules VI and VII, and so do some regions of the medulla, which may provide a copy of eye-movement signals, in support of ocular tracking (Belknap & McCrea 1988). Even some fibers from the sensory nucleus of the trigeminal nerve reach the boundary of the anterior and posterior lobe vermis (lobules V and VI) (Carpenter & Hanna 1961). Lobules VI and VII use this varied information to orient the head and eyes toward interesting objects. They also participate in covert shifts of attention (i.e., attentional shifts that do not involve movement) (Baier *et al.* 2010).

Lobules VII–X of the posterior-lobe vermis receive a large number of spinal afferents, carrying information about body posture and movement. They also have significant input from the pons, relaying signals from the motor and supplementary motor cortices, possibly for voluntary movements such as reaching (Coffman *et al.* 2011). Crus I and II seem to have a wide variety of functions, including some nonmotor functions.

In contrast to the situation in other brain regions such as the cerebral cortex, where sensory and motor information are organized in somatotopic maps, the somatosensory input to the posterior lobe is not somatotopically organized. This may be because a lot of the somatosensory input originates from higher-level sensory areas (via the pons) or from the reticular formation. For example, a single P-cell in the lateral regions of lobules VI and VII (crus II) might receive somatosensory input from both sides of the body and both fore- and hindlimbs as well as the vibrissae (Holtzman *et al.* 2006). Motor representations for somatic musculature in the cerebellum also are not somatotopic, and sensory and motor responses in the same region may be from different body regions (Wiestler *et al.* 2011). This reshuffling of somatosensory and motor information may allow the posterior lobe to learn a wide range of associations. The hemispheres also receive input from some of the same pontine nuclei that project to the known eye movement regions of the cerebellum (the posterior lobe vermis and vestibulocerebellum). These inputs from the pontine nuclei may contain information from several sensory pathways.

The lateral part of lobule IX, also called the "dorsal paraflocculus," is a small but interesting part of the posterior lobe. Like the "oculomotor vermis" (lobules VI and VII), it receives input from visual areas of the

---

[9] The NRTP is in the pontine reticular formation, dorsal to the "pontine nuclei" as normally defined, and its connections and its structure are also distinct from those of the pontine nuclei. The NRTP receives input mostly from the dentate and interpositus nuclei, as described in (Brodal & Szikla 1972). There is also a projection from the vestibular nuclei to the NRTP, as well as a small projection from the cerebral cortex.

brainstem such as the NRTP. It also receives a unique set of climbing fiber inputs that appear to be motor in function, unlike most climbing fibers, which carry sensory signals (Kralj-Hans *et al.* 2007).

## EFFERENT CONNECTIONS OF THE CEREBELLUM

Some of the output of the cerebellum is directed to the medulla and spinal cord. For example, all of the DCN project to the inferior olive (they also project back to the cerebellar cortex (Figure 1.4)). As the cerebral cortex became larger during mammalian evolution, the output of the cerebellum became increasingly directed to the thalamus, which provides a link to the cerebral cortex. This closes the anatomical loop, effectively including the cerebellum in cerebral cortical networks. In general, each region of the cerebellar cortex receives input from the same region of the thalamus to which it provides (indirect) input (Strick *et al.* 2009). So far, the evidence supports the idea that the cerebellum consists of modules that are interconnected with other brain regions in closed loops.

### *Efferents from the vestibulocerebellum*

The vestibulocerebellum has slightly different connections for its medial and lateral parts. Like the anterior-lobe vermis, the vermis of lobules IX and X projects to the vestibular nuclei and fastigial nuclei. Narrow longitudinal zones within IX and X also project to the interpositus and dentate nuclei (Bernard 1987). The flocculus (lobule X) projects almost exclusively to the vestibular nuclei; a small region also projects to the dentate nucleus. Together, the vestibulocerebellum and the fastigial nuclei are mainly responsible for the cerebellum's contribution to perception of self-motion, spatial orientation, and gaze stabilization.

### *Efferents from the anterior lobe*

The anatomical connections of the anterior lobe suggest that its main role is in balance and coordination. It carries out sensory and motor computations for posture and locomotion, which are not simple problems. The anterior-lobe vermis projects to the rostral part of the fastigial nucleus. Other inputs to the fastigial nucleus include direct sensory input from the vestibular nuclei and spinal cord and higher-level input from the red nucleus, a large premotor nucleus in the brainstem. As this arrangement suggests, the fastigial nucleus is both sensory and motor in function.

The rostral fastigial nucleus projects to the lateral vestibular nucleus (also called "Deiters' nucleus"), which contains enormous premotor neurons that project directly to spinal-cord motoneurons. Some of the P-cells of the anterior lobe project directly to Deiters' nucleus, bypassing the fastigial nucleus. Thus the anterior lobe has two output pathways, one direct and one via the fastigial nucleus, that terminate on the large premotor neurons in Deiters' nucleus. There is also a direct projection from the fastigial nucleus to motoneurons of the spinal cord, as well as a projection to the motor cortex (Hashimoto *et al.* 2010).

The more lateral parts of the anterior lobe project somatotopically to the nucleus interpositus, which is similar in its anatomical connections (and probably also its function) to the rostral fastigial nucleus.

## Efferents from the posterior lobe

Lobules VI and VII of the posterior-lobe vermis project to the fastigial oculomotor region (FOR), which is located in the caudal part of the fastigial nucleus. The FOR is tightly linked to the control of eye movements. Lobules VI and VII may also be important for directing attention, which might be the function of the projection from the FOR to the thalamus.

Paravermal regions of the posterior lobe project to the caudal part of the interpositus nuclei, to regions that are involved in the control of limb movements, blinks, gaze shifts, vergence, and accommodation[10] (Zhang & Gamlin 1998). The caudal interpositus projects to regions of the premotor cortex that are involved in planning limb movements (Hashimoto *et al.* 2010). The paravermis of the posterior lobe is also known to be involved in an interesting type of motor learning involving aversive blinks.

The interpositus nuclei have some oculomotor regions, and these receive input from the lateral part of lobule IX (the dorsal paraflocculus), one of several oculomotor regions of the cerebellum. The dorsal paraflocculus is also thought to be involved in eye–hand coordination, or visually guided reaching as it is more accurately called (Kralj-Hans *et al.* 2007).

The enormous hemispheres of the human posterior lobe project primarily to the (also enormous) dentate nucleus. In turn, the dentate nucleus has a major projection to the thalamus, effectively connecting the cerebellum back to the cerebral cortex in a closed loop. The dentate nucleus is involved in a multitude of functions. For a start, it projects in somatotopic fashion, via the thalamus, to the primary motor cortex. This projection supports the role of the cerebellar hemispheres in coordinating skilled movements of the hands, arms, and legs. Specialized subregions of the dentate nucleus are involved in tracking eye movements and in blinks. Interestingly, the dentate nucleus also projects (indirectly) to regions involved in movement planning, in the premotor cortex (Hashimoto *et al.* 2010).

[10]Vergence eye movements and accommodation of the lens and pupil allow us to focus on near visual targets.

In addition to its motor functions, the dentate nucleus of primates has taken on some cognitive tasks, perhaps accounting for its extra-large size. It is connected (via the thalamus) to several nonmotor cortical regions, including a region of the dorsolateral prefrontal cerebral cortex that is essential for spatial working memory and motor planning (Middleton & Strick 1994). The dentate also projects to a region of the posterior parietal cortex that is thought to be involved in perceiving the location of the head and/or body, and other objects, in space (Clower *et al.* 2001). These different projections arise from separate regions of the dentate nucleus. Overall, roughly 40–50% of the dentate is believed to project to nonmotor areas, even in monkeys (Dum & Strick 2003); in humans the dentate is relatively larger than it is in monkeys. In summary, then, its connectivity suggests that the cerebellum contributes to nonmotor, as well as motor aspects of behavior.

## REFERENCES


Apps, R. (2000) Rostrocaudal branching within the climbing fibre projection to forelimb-receiving areas of the cerebellar cortical C1 zone. *J. Comp. Neurol.*, **419**, 193–204.

Apps, R. & Hawkes, R. (2009) Cerebellar cortical organization: a one-map hypothesis. *Nature Rev. Neurosci.*, **10**, 670–681.

Baier, B., Dieterich, M., Stoeter, P., Birklein, F. & Muller, N.G. (2010) Anatomical correlate of impaired covert visual attentional processes in patients with cerebellar lesions. *J. Neurosci.*, **30**, 3770–3776.

Barmack, N.H. & Yakhnitsa, V. (2003) Cerebellar climbing fibers modulate simple spikes in Purkinje cells. *J. Neurosci.*, **23**, 7904–7916.

Barmack, N.H. & Yakhnitsa, V. (2008) Functions of interneurons in mouse cerebellum. *J. Neurosci.*, **28**, 1140–1152.

Barmack, N.H. & Yakhnitsa, V. (2011) Microlesions of the inferior olive reduce vestibular modulation of Purkinje cell complex and simple spikes in mouse cerebellum. *J. Neurosci.*, **31**, 9824–9835.

Belknap, D.B. & McCrea, R.A. (1988) Anatomical connections of the prepositus and abducens nuclei in the squirrel monkey. *J. Comp. Neurol.*, **268**, 13–28.

Bengtsson, F., Svensson, P. & Hesslow, G. (2004) Feedback control of Purkinje cell activity by the cerebello-olivary pathway. *Eur. J. Neurosci.*, **20**, 2999–3005.

Bernard, J.F. (1987) Topographical organization of olivocerebellar and corticonuclear connections in the rat–An WGA-HRP study: I. Lobules IX, X, and the flocculus. *J. Comp. Neurol.*, **263**, 241–258.

Braitenberg, V. & Atwood, R.P. (1958) Morphological observations on the cerebellar cortex. *J. Comp. Neurol.*, **109**, 1–33.

Brodal, A. & Szikla, G. (1972) The termination of the brachium conjunctivum descendens in the nucleus reticularis tegmenti pontis. An experimental anatomical study in the cat. *Brain Res.*, **39**, 337–351.

Brodal, A. (1981) *Neurological Anatomy in Relation to Clinical Medicine*, Oxford University Press, New York.

Brodal, P. (1978) The corticopontine projection in the rhesus monkey. Origin and principles of organization. *Brain*, **101**, 251–283.

Brown, M.E. & Ariel, M. (2009) Topography and response timing of intact cerebellum stained with absorbance voltage-sensitive dye. *J. Neurophysiol.*, **101**, 474–490.

Carpenter, M.B. & Hanna, G.R. (1961) Fiber projections from the spinal trigeminal nucleus in the cat. *J. Comp. Neurol.*, **117**, 117–131.

Carpenter, M.B. (1976) *Human Neuroanatomy*, Williams & Wilkins, Baltimore.

Clower, D.M., West, R.A., Lynch, J.C. & Strick, P.L. (2001) The inferior parietal lobule is the target of output from the superior colliculus, hippocampus, and cerebellum. *J. Neurosci.*, **21**, 6283–6291.

Coffman, K.A., Dum, R.P. & Strick, P.L. (2011) Cerebellar vermis is a target of projections from the motor areas in the cerebral cortex. *Proc. Natl. Acad. Sci. USA*, **108**, 16068–16073.

Cohen, D. & Yarom, Y. (1998) Patches of synchronized activity in the cerebellar cortex evoked by mossy-fiber stimulation: questioning the role of parallel fibers. *Proc. Natl. Acad. Sci. USA*, **95**, 15032–15036.

Cohen, D. & Yarom, Y. (2000) Cerebellar on-beam and lateral inhibition: two functionally distinct circuits. *J. Neurophysiol.*, **83**, 1932–1940.

de Solages, C., Szapiro, G., Brunel, N., Hakim, V., Isope, P., Buisseret, P., Rousseau, C., Barbour, B. & Lena, C. (2008) High-frequency organization and synchrony of activity in the Purkinje cell layer of the cerebellum. *Neuron*, **58**, 775–788.

De Zeeuw, C.I., Hoebeek, F.E., Bosman, L.W.J., Schonewille, M., Witter, L. & Koekkoek, S.K.E. (2011) Spatiotemporal firing patterns in the cerebellum. *Nature Rev. Neurosci.*, **12**, 327–344.

DiGregorio, D.A. & Nusser, Z. (2002) Spillover of glutamate onto synaptic AMPA receptors enhances fast transmission at a cerebellar synapse. *Neuron*, **35**, 521–533.

Dum, R.P. & Strick, P.L. (2003) An unfolded map of the cerebellar dentate nucleus and its projections to the cerebral cortex. *J. Neurophysiol.*, **89**, 634–649.

Eccles, J.C., Ito, M. & Szentagothai, J. (1967) *The Cerebellum as a Neuronal Machine*, Springer, Berlin.

Engbers, J.D.T., Anderson, D., Asmara, H., Rehak, R., Mehaffey, W.H., Hameed, S., McKay, B.E., Kruskic, M., Zamponi, G.W. & Turner, R.W. (2012) Intermediate conductance calcium-activated potassium channels modulate summation of parallel fiber input in cerebellar Purkinje cells. *Proc. Natl. Acad. Sci. USA*, **109**, 2601–2606.

Gao, W., Chen, G., Reinert, K.C. & Ebner, T. (2006) Cerebellar cortical molecular layer inhibition is organized in parasagittal zones. *J. Neurosci.*, **26**, 8377–8387.

Gauck, V. & Jaeger, D. (2000) The control of rate and timing of spikes in the deep cerebellar nuclei by inhibition. *J. Neurosci.*, **20**, 3006–3016.

Graham, D.J. & Wylie, D.R. (2012) Zebrin-immunopositive and -immunonegative stripe pairs represent functional units in the pigeon vestibulocerebellum. *J. Neurosci.*, **32**, 12769–12779.

Grant, G. (1962) Projection of the external cuneate nucleus onto the cerebellum in the cat: an experimental study using silver methods. *Exp. Neurol.*, **5**, 179–195.

Groenewegen, H.J., Voogd, J. & Freedman, S.L. (1979) The parasagittal zonal organization within the olivocerebellar projection. *J. Comp. Neurol.*, **183**, 551–602.

Hashimoto, M., Takahara, D., Hirata, Y., Inoue, K., Miyachi, S., Nambu, A., Tanji, J., Takada, M. & Hoshi, E. (2010) Motor and non-motor projections from the cerebellum to rostrocaudally distinct sectors of the dorsal premotor cortex in macaques. *Eur. J. Neurosci.*, **31**, 1402–1413.

Holtzman, T., Rajapaksa, T., Mostofi, A. & Edgley, S.A. (2006) Different responses of rat cerebellar Purkinje cells and Golgi cells evoked by widespread convergent sensory inputs. *J. Physiol.*, **574**, 491–507.

Hounsgaard, J. & Yamamoto, C. (1979) Dendritic spikes in Purkinje cells of the guinea pig cerebellum studied in vitro. *Exp. Brain Res.*, **37**, 387–398.

Jones, E.G. & Wise, S.P. (1978) Size, laminar and columnar distribution of efferent cells in the sensory-motor cortex of monkeys. *J. Comp. Neurol.*, **175**, 391–438.

Kimpo, R. & Raymond, J.L. (2007) Impaired motor learning in the vestibulo-ocular reflex in mice with multiple climbing fiber input to cerebellar Purkinje cells. *J. Neurosci.*, **27**, 5672–5682.

Kralj-Hans, I., Baizer, J.S., Swales, C. & Glickstein, M. (2007) Independent roles for the dorsal paraflocculus and vermal lobule VII of the cerebellum in visuomotor coordination. *Exp. Brain Res.*, **177**, 209–222.

Langer, T., Fuchs, A.F., Scudder, C.A. & Chubb, M.C. (1985) Afferents to the flocculus of the cerebellum in the rhesus macaque as revealed by retrograde transport of horseradish peroxidase. *J. Comp. Neurol.*, **235**, 1–15.

McKay, B.E., Engbers, J.D.T., Mehaffey, W.H., Gordon, G.R.J., Molineux, M.L., Bains, J.S. & Turner, R.W. (2007) Climbing fiber discharge regulates cerebellar functions by controlling the intrinsic characteristics of Purkinje cell output. *J. Neurophysiol.*, **97**, 2590–2604.

Middleton, F.A. & Strick, P.L. (1994) Anatomical evidence for cerebellar and basal ganglia involvement in higher cognitive function. *Science*, **266**, 458–461.

Miles, F.A., Fuller, J.H., Braitman, D.J. & Dow, B.M. (1980) Long-term adaptive changes in primate vestibuloocular reflex. III. Electrophysiological observations in flocculus of normal monkeys. *J. Neurophysiol.*, **43**, 1437–1476.

Mitchell, S.J. & Silver, A. (2000) Glutamate spillover suppresses inhibition by activating presynaptic mGluRs. *Nature*, **404**, 498–502.

Montarolo, P.G., Palestini, M. & Strata, P. (1982) The inhibitory effect of the olivocerebellar input on the cerebellar Purkinje cells in the rat. *J. Physiol.*, **332**, 187–202.

Mugnaini, E. & Floris, A. (1994) The unipolar brush cell: a neglected neuron of the mammalian cerebellar cortex. *J. Comp. Neurol.*, **339**, 174–180.

Nagao, S., Kitamura, T., Nakamura, N., Hiramatsu, T. & Yamada, J. (2002) Differences of the primate flocculus and ventral paraflocculus in the mossy and climbing fiber input organization. *J. Comp. Neurol.*, **382**, 480–498.

Nishiyama, H. & Linden, D.J. (2004) Differential maturation of climbing fiber innervation in cerebellar vermis. *J. Neurosci.*, **24**, 3926–3932.

Osanai, R., Nagao, S., Kitamura, R., Kawabata, I. & Yamada, J. (1999) Differences in mossy and climbing afferent sources between flocculus and ventral and dorsal paraflocculus in the rat. *Exp. Brain Res.*, **124**, 248–264.

Oscarsson, O. (1979) Functional units of the cerebellum: sagittal zones and microzones. *Trends Neurosci.*, **2**, 144–145.

Palkovits, M., Mezey, E., Hamori, J. & Szenthagothai, J. (1977) Quantitative histological analysis of the cerebellar nuclei in the cat. I. Numerical data on cells and on synapses. *Exp. Brain Res.*, **28**, 189–209.

Raman, I.M. & Bean, B.P. (1997) Resurgent sodium current and action potential formation in dissociated cerebellar Purkinje neurons. *J. Neurosci.*, **17**, 4517–4526.

Raman, I.M., Gustafson, A.E. & Padgett, D. (2000) Ionic currents and spontaneous firing in neurons isolated from the cerebellar nuclei. *J. Neurosci.*, **20**, 9004–9016.

Rambold, H., Churchland, A., Selig, Y., Jasmin, L. & Lisberger, S.G. (2002) Partial ablations of the flocculus and ventral paraflocculus in monkeys cause linked deficits in smooth pursuit eye movements and adaptive modification of the VOR. *J. Neurophysiol.*, **87**, 912–924.

Rokni, D., Tal, A., Byk, H. & Yarom, Y. (2009) Regularity, variability and bistability in the activity of cerebellar Purkinje cells. *Front Cell Neurosci.*, **3**, 1–9.

Russo, M.J., Yau, H.-J., Nunzi, M.G., Mugnaini, E. & Martina, M. (2008) Dynamic metabotropic control of intrinsic firing in cerebellar unipolar brush cells. *J. Neurophysiol.*, **100**, 3351–3360.

Sarnat, H.B. & Netsky, M.G. (1974) *Evolution of the Nervous System*, Oxford University Press, New York.

Strick, P.L., Dum, R.P. & Fiez, J.A. (2009) Cerebellum and nonmotor function. *Ann. Rev. Neurosci.*, **32**, 413–434.

Telgkamp, P. & Raman, I.M. (2002) Depression of inhibitory synaptic transmission between Purkinje cells and neurons of the cerebellar nuclei. *J. Neurosci.*, **22**, 8447–8457.

van der Giessen, R.S., Maxeiner, S., French, P.J., Willecke, K. & de Zeeuw, C.I. (2006) Spatiotemporal distribution of Connexin45 in the olivocerebellar system. *J. Comp. Neurol.*, **495**, 173–184.

Voogd, J. & Glickstein, M. (1998) The anatomy of the cerebellum. *Trends Neurosci.*, **21**, 370–375.

Voogd, J., Pardoe, J., Ruigrok, R.J. & Apps, R. (2003) The distribution of climbing and mossy fiber collateral branches from the copula pyramis and the paramedian lobule: congruence of climbing fiber cortical zones and the pattern of zebrin banding within the rat cerebellum. *J. Neurosci.*, **23**, 4645–4656.

Wang, X., Chen, G., Gao, W. & Ebner, T. (2011) Parasagittally aligned, mGluR1-dependent patches are evoked at long latencies by parallel fiber stimulation in the mouse cerebellar cortex in vivo. *J. Neurophysiol.*, **105**, 1732–1746.

Wiestler, T., McGonigle, D.J. & Diedrichsen, J. (2011) Integration of sensory and motor representations of single fingers in the human cerebellum. *J. Neurophysiol.*, **105**, 3042–3053.

Zhang, H. & Gamlin, P.D.R. (1998) Neurons in the posterior interposed nucleus of the cerebellum related to vergence and accomodation. *J. Neurophysiol.*, **79**, 1255–1269.

## FURTHER READING

Brodal, A. (1981) *Neurological Anatomy in Relation to Clinical Medicine*, Oxford University Press, New York; Chapter 5.

Eccles, J.C., Ito, M. & Szentagothai, J. (1967) *The Cerebellum as a Neuronal Machine*, Springer, Berlin.

Sarnat, H.B. & Netsky, M.G. (1974) *Evolution of the Nervous System*, Oxford University Press, New York; Chapters 5 and 6.

Voogd, J. & Glickstein, M. (1998) The anatomy of the cerebellum. *Trends Neurosci.*, **21**, 370–375.

# 2 Operating the Machine

Now that we have a basic understanding of the structure of the cerebellum, we can ask some very general questions about how it might operate. The possible answers that I will present arose from theory and imagination, but seem likely to be correct. In this chapter, I will mostly summarize the ideas. Experimental evidence supporting them will be presented later on in the book.

As shown in Figure 1.1, the anatomy of the cerebellar cortex suggests a data processor, with its very regular, rectangular lattice structure. In the sagittal plane, each Purkinje cell (P-cell) dendritic arbor occupies a large, roughly rectangular area, a few hundred microns on a side. However, in the coronal or transverse plane, the P-cell dendritic arbors are fairly thin, and they are lined up like dominoes. Meanwhile, the parallel fibers extend in the coronal plane (perpendicular to the dominoes) for up to 6.6 mm (Mugnaini 1983). This arrangement maximizes the number of P-cells that each parallel fiber can contact. In fact, each one terminates on roughly 100 P-cells, and at least 100,000 parallel fibers synapse on each P-cell (in cats) (Eccles 1977). Because of this arrangement, each P-cell has a very large contingent of parallel fibers that are shared among a row of P-cells, running along the folium in the transverse plane.

These features have implications for what the cerebellum can do. For example, the arrangement of parallel fibers and P-cells seems perfect for timing, a function that requires rapid, precise calculation. In fact, the cerebellum does participate in timing of movements. But even before this was known, Valentino Braitenberg did some calculations based on cerebellar cortical structure and on his guess that the cerebellar cortex might be a timekeeping device. One of Braitenberg's assumptions was that some cerebellar neurons fire action potentials synchronously (Braitenberg 1961).[1] Later, his ideas were supported when it turned out

[1]Braitenberg's original hypothesis had three components, of which I am discussing only two here. I have skipped the famous delay line idea because Braitenberg himself tossed it out later.

*The Cerebellum: Learning Movement, Language, and Social Skills*, First Edition. Dianne M. Broussard.

that synchronous firing is an important feature of the cerebellar circuit (Llinas & Sasaki 1989).

One of Braitenberg's hypotheses is the "stopwatch" function. The idea is that a group of P-cells can measure time intervals by using the synchronous firing of climbing fibers as a reference point. In this scenario, each complex spike would coincide with input from a particular group of mossy fibers, which are a particular distance away from the P-cell. As a result, the information provided by those particular mossy fibers would receive some special kind of processing. Another, similar idea based on the convergence of climbing-fiber and mossy-fiber inputs is that two functions of time (which could represent some real-world variable) would be cross-correlated by the cerebellar cortex (Braitenberg 1961). The idea behind cross-correlation is as follows. Suppose you find a key in your junk drawer; you do not know what this key will unlock, but one way to find out quickly is to compare it with the other keys in your possession. You hold your unknown key up to each of the other keys, align them as closely as you can, and compare all of the fine features. Usually in cross-correlation, the long axis of your key would become time (although it does not have to). Also, when you aligned your keys you optimized the "time delay," but in cross-correlation you do not bother to do that; you just add up the results for all of the possible time delays.

These two ideas, along with a few others, became known as the cerebellar clock hypothesis. Although Braitenberg's guess was more or less correct, other modelers did not adopt it until quite recently. One reason for this was that cerebellar theoreticians became obsessed with the problem of learning.

## LEARNING IN THE CEREBELLAR CORTEX

If cerebellar activity somehow encodes movement components, using the stopwatch or a similar function, the cerebellum also might be able to assemble more complex stereotyped movements, such as an accurate throw. By analogy to computers, the representations in the cerebellum of these movements were given the name of "motor programs" (long before there was any evidence that movement plans actually exist). In 1969, David Marr proposed that the cerebellum stores motor programs for later use (Marr 1969). Marr was a 24-year-old graduate student at Cambridge when he published his famous paper (sadly, he died of leukemia at only 35 years of age.)

The central hypothesis of Marr's thought experiment was that each P-cell, along with its associated climbing fiber, encodes a single component of the program. A component might be rotation around a single joint. When the climbing fiber fires, it indicates that the movement component should be carried out, and the P-cell fires. Meanwhile, the parallel-fiber inputs to the same P-cell provide the *context* that becomes associated with that elemental movement. Marr suggested that the P-cell learns to generate its elemental movement when it is exposed to the correct context, even without the climbing-fiber input. He estimated that P-cells might learn to recognize hundreds or thousands of context "events," and any one of these would be able to bring about the correct motor response.

A similar theory was proposed by James Albus, who was working for the National Aeronautics and Space Administration (NASA) at the time and did not even know of Marr's theory. Albus pointed out that the structure of the parallel fibers and P-cell dendritic arbors are arranged in a way that maximizes both convergence[2] and divergence and, hence, the information capacity of each module (Albus 1971).

Many of the ideas from these early models have stood the test of time. One especially useful idea was that cerebellar learning would be guided by trial and error. Marr proposed that, in some cases, the climbing fibers would encode some quantity that is minimized by learning: an "error signal." The error signal would be the difference between what you intended to do, and what you actually did; say, an error in the position of your hand while reaching. Albus had the same idea but carried it farther, taking advantage of the observation that the climbing fiber input to P-cells causes a pause in the steady stream of simple spikes (Granit & Phillips 1956). Albus suggested that we think of the pause as an unconditioned response (borrowing the term from associative-learning theory). Then we can relate the firing patterns of cerebellar neurons during learning to the real-world events that actually bring about learning. For example, in the flocculus, the pause can be brought about by a visual stimulus that represents an inaccurate vestibulo-ocular reflex.

Building on the analogy to behavioral conditioning, Albus next suggested that the mossy-fiber inputs represented conditioned stimuli. In other words, after learning, the mossy-fiber input alone should elicit exactly the same response from P-cells that was originally elicited by the climbing fiber. This idea is similar to the context hypothesis of Marr. Since the P-cell response to the climbing fiber was a pause in firing, it followed that the excitatory input from parallel fibers to P-cells should be *weakened* by learning. Then, a steady stream of action potentials arriving via mossy fibers would excite the P-cell much less than it did before learning occurred. Importantly, Albus suggested that if the P-cell were to fire an action potential immediately after the climbing-fiber input, every synapse associated with generating that action potential should be weakened, or "depressed." Albus was right; this pattern of activation does decrease the

[2]Convergence is the number of presynaptic elements that provide input to the postsynaptic neuron. Divergence is the number of postsynaptic elements that receive input from a neuron.

effect of the mossy-fiber input on the P-cells (Sakurai 1987). The decrease is cerebellar long-term depression (LTD).

In Albus' theory, the information capacity of the network would be greatly reduced if synaptic weights were *only* permitted to depress, and not to increase (or "potentiate") during learning. So, he next proposed that both potentiation and depression are possible, depending on the input signals. This was correct too; researchers eventually demonstrated cerebellar long-term potentiation (LTP) (Salin *et al.* 1996). But for learning, LTD has the intrinsic advantage that synaptic strengths tend to decrease asymptotically. In other words, the more a synapse has been weakened, the less effect the convergent inputs will have, and the less likely it is that more weakening will occur. Synaptic weights can never reach zero via LTD. With LTP alone, in fact, they quickly saturate; in other words, they reach their physical limits. Then, some mechanism of decay of the potentiated synaptic weights becomes necessary. Depression is intrinsically more stable. It seems to follow that this advantage of LTD might be the reason for the persistence of cerebellar memories over long periods of disuse, as no decay mechanism is required. Furthermore, the stability of LTD might explain why motor learning is successful with only intermittent training.

## PATTERN RECOGNITION BY THE CEREBELLUM

Another influential idea proposed by both Marr and Albus is that the cerebellar cortex operates as a pattern recognition device. This idea is somewhat similar to Braitenberg's cross-correlation idea. Interestingly, Marr proposed that the Golgi cells participate in pattern recognition, because they regulate the sizes of groups of mossy fibers that can activate particular granule cells (the "codon size"). Similarly, in Albus' theory, the role of Golgi cells is to keep the number of parallel fibers that are excited by a given input signal relatively constant, so that they do not depend on irrelevant characteristics of the input (such as intensity or bandwidth).

For pattern recognition, which can require a lot of computation, Albus proposed that the cerebellum carries out "expansion recoding." Expansion recoding is coding of information in greater detail. This would be a good use of the storage capacity of the cerebellum, and recognition of the pattern can be honed to greater precision by learning. The granule cells are the elements of the expansion recoder. The enormous number of granule cells would allow us to learn to recognize a large number of sensory or motor input signals, provided that these arrive independently via mossy fibers. Because there are so many of them, individual granule cells would have distinct patterns of activity with little or no concurrent activation, and hence very little ambiguity. Because of the low ambiguity, expansion recoding permits rapid, efficient learning of discriminations among events, or input patterns. Thus, Albus' model takes advantage of the estimated 150,000 parallel-fiber inputs to each P-cell to enormously expand the information capacity of the cerebellar cortex.

Albus' model also suggested a function for stellate and basket cells in extending the possible weights of the inputs to P-cells into the negative range (because these interneurons are inhibitory). Once these neurons were incorporated into the model, there was an obvious advantage of synaptic plasticity at the synapses between parallel fibers and the stellate and basket cells, that is, the negative weights could be adjusted. Albus' prediction that the inputs to these interneurons must be plastic has since received experimental support (Jorntell & Ekerot 2002).

After Marr and Albus published their papers, there was a consensus that the function of the cerebellum was likely to be learning. Masao Ito pointed out the potential usefulness of cerebellar learning for controlling feed-forward systems like the vestibulo-ocular reflex (Ito 1972). A feed-forward system has no built-in way of keeping track of its own performance; all of its connections proceed in one direction, like a river. This is in contrast to feedback systems. Feed-forward systems absolutely must be able to learn in order to perform adequately.

Based on the large number of gap junctions in the inferior olive, and the synchrony of some climbing-fiber discharges, Rodolfo Llinas proposed that climbing fibers synchronize groups of P-cells and their target neurons in the deep nuclei cerebellar (DCN), linking them functionally into groups that subserve elemental movements, not unlike those in Marr's model. Experimental evidence for this idea was provided later (Llinas & Sasaki 1989; Welsh *et al.* 1995; Lang *et al.* 1999). Llinas also promoted the "tensor model" of cerebellar function, which provides a method by which the cerebellum could convert sensory information into motor coordinates (Pellionisz & Llinas 1979). This conversion is a problem that the brain has to solve somehow, because sensory and motor systems use reference frames and coordinate axes that are fundamentally different. The idea in the tensor model was that sensory systems use "covariant" coordinates, which represent information as the vector sum of orthogonal components that are derived from the sense organs. Meanwhile, motor systems use contravariant vectors. They represent information as a vector sum, but the components need not be orthogonal (e.g., they could represent individual muscles or joints).

Llinas did not agree that the cerebellum was a learning device, but thought of it as a timing and computational device. Since the 1970s, the overwhelming bulk of the evidence has favored the idea that the cerebellum is a learning machine; however, one of the things that it learns is timing. The evidence also shows that synchrony has an important part in learning, in the expression of memory, in pattern recognition, and in timing of some aspects of behavior.

## NEURAL NETWORKS

Neurons integrate information by combining their synaptic inputs and generating an action potential if the membrane potential at the spike generator exceeds threshold. However, synaptic inputs may contribute differently depending on where they are located, how many receptor molecules they contain, and other factors. Because a neuron is so complicated, the best way to guess what a neuron will do (short of studying it directly) is to simulate it mathematically. Such simulated neurons can then be put together into a neural-network model. For example, Albus' model is a neural network, in which individual neurons are represented by equations. One set of variables in these equations represents synaptic "weight," a coefficient that is applied by each synapse. To simulate learning, Albus changed the synaptic weights.

A bit later, Terrence Sejnowski made another neural-network model of the cerebellum. Sejnowski's point was that the cerebellum can be thought of as an adaptive filter (Sejnowski 1977). A filter is selective for certain signals, and in an adaptive filter, this selectivity can be changed. Sejnowski also showed that the properties of the cerebellar filter depends on the background firing rates of its neurons, providing the first glimmer of understanding of why cerebellar neurons need to be spontaneously active.

Although the background firing rates of cerebellar neurons are important, this variable was ignored by many modelers, who assumed that it did not matter. In these models, the cerebellum was assumed to be a linear system. Linear signal processing requires a system whose output is a fixed function of the input, such that the plot of output versus input amplitude is always a straight line. In a linear system, the relationship of output to input is *constant*. This rule really can apply only within a finite range, because eventually any system will reach a limit in the output it can generate. Therefore, we say that systems are linear *within a range* of input signals.

Why did modelers assume that the cerebellum was linear? Well, it is true that most of the time, cerebellar neurons fire at fairly steady rates that can be represented by continuous functions of time (i.e., they do not start, stop, or jump to a new value). Also, using a linear model we might find out a lot in a short time, because some linear models are relatively simple.

## THE CEREBELLUM AS PART OF A "CONTROL SYSTEM"

Control systems models are (usually) linear models that describe the response of systems *as they change over time*. Usually the independent variable is frequency rather than time, greatly simplifying the mathematics.[3] Linear control systems models have great explanatory power wherever rate of change is a variable, as obviously it is for motor control. Starting in the 1960s, engineers used control systems theory to create models of motor systems. David Robinson (Robinson 1976), Roger Carpenter (Carpenter 1972) and others incorporated the cerebellum into their models and assigned it particular frequency dependence or filtering, characteristics that could affect the speed and acceleration of movements.

Combining control systems theory with neural network theory, Fujita developed a model that would simulate cerebellar motor learning as optimization of a linear filter (Fujita 1982). A linear filter is a linear system that is selective for a particular kind of signal; for example, a low-pass filter will selectively transmit low frequencies. Fujita's model was meant to apply to the vestibulocerebellum, which may have a larger linear range than other cerebellar regions (Kim *et al.* 2012). In the model, the cerebellum does two things. First, it functions as a phase lead or lag[4] compensator (this could adjust the timing of movements, although not very precisely). Second, it learns under the guidance of error signals.

Like other signals, the error signal is a function of time, and in linear models it is a continuous, ongoing function. An error signal in real life could be obtained from a sensory input. For example, let's look at the vestibulo-ocular reflex. Its job is to stabilize the eyes with respect to the outside world, so we can see clearly. Vision can report whether or not the eyes are stable. So in the vestibulo-ocular reflex, the error signal can be obtained by vision. The learning algorithm in Fujita's model makes changes in synaptic weights depending on the error signal. This may be the way that motor learning actually works.

Although Fujita's model was developed for motor learning, it may have implications for other types of learning as well. As we will see at the end of this book, some of the functions of the cerebellum may involve generating imitations of cerebral cortical systems that can then be perfected and used to react more rapidly to a very broad range of situations, including (for example) social ones. In this broader context, error signals may represent the mismatch between the cerebellar output and the cortical output that it will eventually imitate. Exactly what is accomplished

---

[3]In control systems theory, systems are represented by differential equations or more simply by Laplace transforms that predict their response to perturbations. Physical properties (such as mass) of the system described are independent variables in these simple equations. Laplace transforms predict responses in terms of the frequency of the perturbation (i.e., cycles per second).

[4]Phase here refers to the phase angle of simple harmonic motion (e.g., a sine wave). A phase lag means that the peak amplitude of the output signal is delayed with respect to the input of the system; phase lead means that peak output precedes the peak in the input signal.

by a particular region of the cerebellar cortex will depend on its input and output connections, but the basic circuitry is the same. However, some error signals are not continuous, and the systems controlling them may be fundamentally nonlinear, so the cerebellum may handle them differently.

## MULTIPLE SITES FOR CEREBELLAR LEARNING?

In addition to potentiation and depression at the parallel-fiber inputs to P-cells and in the deep nuclei, Albus' model suggested that long-term changes in the inputs to inhibitory interneurons would be helpful. In fact, it may be a general rule, rather than the exception, that synapses are capable of long-term plasticity.

Another control systems model of motor learning, this time incorporating two sites for plasticity, was advanced by Steve Lisberger (Lisberger 1994). Lisberger used some of the more general characteristics of real cerebellar neurons for his model of the vestibulo-ocular reflex. He found that he could simulate learning by making adjustments at two loci, one at the parallel-fiber inputs to P-cells, and one at the mossy-fiber input to a vestibular neuron (analogous to a DCN neuron). In fact, if learning occurred only at the DCN neuron, the reflex could not be adjusted enough to match the experimental data. But if learning occurred only at the inputs to P-cells, the reflex became unstable.

To understand this problem, we need to spend a few moments thinking about how movements are controlled. Smooth pursuit eye movements, which track visual objects, must be able to continue when the eyes are perfectly on target and the image of the target is perfectly steady on the retina. But when this happens, there is no visual signal that can guide pursuit. In a linear system, if the input signal falls to zero, so does the output signal; we need a substitute signal from somewhere. In Lisberger's model, which uses the configuration developed by Fred Miles (Miles & Eighmy 1980), the motor output of the cerebellum contributes to a positive-feedback loop. An eye-velocity signal is fed back from a premotor center to provide the signal to keep the eyes moving even when perfectly on target. Miles and Lisberger proposed that the smooth-pursuit feedback loop incorporates some of the same P-cells that are also involved in the vestibulo-ocular reflex. Of course, if positive feedback is present, changes in the motor output could build on themselves and, potentially, cause signals to go out of control (think of what happens when a microphone is placed too close to a speaker). To avoid this problem, Lisberger proposed that learning must occur simultaneously at both sites: the brainstem site, and the P-cell.

In the 1990s, Javier Medina, Nelson Donegan, and Michael Mauk followed up on Lisberger's idea and showed its feasibility using a simulated neural network. Unlike the control systems models, Mauk's models are not required to be linear, and they can learn without human supervision. The networks are very large, and follow known principles of cerebellar anatomy, such as the actual convergence and divergence ratios. The idea is to give them realistic inputs, and see what happens; this is the "bottom up" approach to modeling (Medina & Mauk 2000).

Mauk's networks incorporate a simplified version of neuronal membrane properties, including a firing threshold and some conductances; a large number of granule cells (about 10,000); and proportional numbers of other cerebellar cell types. The network contains over 300,000 synapses whose weights are adjusted by a computer algorithm while the network "learns." Using this kind of model, Mauk's group illustrated how plasticity at the synapses between collaterals of mossy fibers and neurons in the DCN could contribute to learning (Mauk & Donegan 1997; Medina & Mauk 1999). In their model, long-term synaptic plasticity is first induced in the cortex and, later, in the deep nuclei. This happens because as a result of LTD in the cerebellar cortex, P-cell firing is reduced, and this directly results in LTP in the deep nuclei.

At around the same time, Garrett Kenyon pointed out that if cerebellar learning were limited to the parallel-fiber inputs to P-cells, the use of overlapping populations of P-cells to learn large numbers of associations would eventually degrade previously stored memories (Kenyon 1997). At other sites in the circuit, memory may be safer from degradation; for example, plasticity at the excitatory inputs to the stellate and basket cells might provide a secondary mechanism for learned pauses in P-cell activity that are thought to be important for some motor responses.[5] Suppose that, as Mauk proposed, learning first takes place at the parallel fiber–P-cell synapse. Suppose that next, the memory is shifted to the parallel-fiber inputs to inhibitory interneurons instead of being shifted to the DCN. Then, depression of the P-cell's own excitatory input would gradually be *replaced* by increased inhibition.

Recently, a neural-network model of the cerebellum by Paul Dean explained how the cerebellum can compensate for the expected contamination of sensory by motor signals (Dean *et al.* 2002). This problem might seem trivial, but it is one of those situations where poor precision can result in severe consequences (such as colliding with objects). Here is how it works: every time we generate a movement, the movement will be detected by some of our sensory systems. We do not want these extra signals to trigger our stabilizing reflexes, or to interfere with tracking of external

[5]The pause in firing, of course, means the signal is not continuous; it drops to zero for a moment. Simple linear models would not be able to deal with the pause, but Kenyon's model was nonlinear.

objects. In Dean's model, the idea is that the cerebellum generates a prediction of the sensory input that is expected to result from planned movements. The expected sensory input is subtracted from the actual sensory input. The difference, then, represents the *unexpected* component of sensory information.

Importantly, Dean's model made use of expansion recoding within an adaptive filter. This particular adaptive filter has many channels with different dynamic properties, and the weights of the channels can be adjusted independently of each other. In a later version of the model, Dean, along with Henrik Jorntell and others, suggested that parasagittal microzones of cerebellar cortex (which they estimate could number around 5000) are the actual units that encode components of movements (Dean *et al.* 2010). These microzones could correspond to channels of the adaptive filter.

In each of the simulated networks, the synaptic weights require a "learning rule" in order to change automatically. In Dean's model, the rule depends on timing. The idea is simple: if the complex spike precedes the simple spike, LTD occurs; but if the simple spike happens first, LTP occurs. Later on, I will describe the experimental evidence supporting this proposal.

## THE CEREBELLAR CLOCK

Although Braitenberg proposed that the cerebellum might time events, this idea was almost completely ignored for a long time, largely because the linear models could not handle it. The problem is that discrete events start and stop all at once, and the timing of such events cannot be represented very precisely by linear models. But in real life, any organism that ignored the timing of discrete events would certainly be eaten by predators. Furthermore, there is convincing evidence that the cerebellum is responsible for timing of movement components as part of a motor "program," that is, a pattern of activity that implements a particular action, and that is used over and over. The movement components could be represented in the cerebellum as mossy-fiber inputs that are processed by something resembling Braitenberg's "stopwatch" algorithm. The stopwatch will filter out representations that are incorrectly timed with respect to the complex spike.

Luckily, an extremely productive group of scientists, led by Chris de Zeeuw and others, is rapidly filling in the gaps in our knowledge of cerebellar timing, and delineating the contributions of the different cell populations in the cerebellar cortex. Other groups have also made tremendous progress on this front in the last few years. This section will summarize some of the major conclusions about the mechanics and function of the cerebellar clock.

To understand cerebellar timing it is necessary to represent neurons even more realistically, by using compartmental models. A compartmental model is a biophysical model in which a neuron is represented by a series of compartments. Membrane properties, ion channels, ion concentrations, and other variables are included in a compartmental model. Using a simulated P-cell in a detailed compartmental model, a recent study by Volker Steuber, Wolfgang Mittman, and others returned to Marr and Albus' pattern-recognition idea. Like real P-cells, the P-cells in the model are spontaneously active. Because their responses are superimposed on a baseline firing rate, the most efficient way to encode information in such cells may be in the duration of pauses in firing, rather than in the number of evoked spikes (Steuber *et al.* 2007).

In the Steuber–Mittman model, the identity of the pattern that arrives at the cerebellar cortex via mossy fibers is encoded in the duration of a pause in P-cell firing that follows the response to the pattern itself. The pause is due to calcium influx into the P-cell during the presentation of the pattern, and persistent activation of calcium-sensitive potassium channels.[6] As the pattern becomes familiar, the pause becomes shorter and shorter, because LTD reduces the amount of calcium influx. The P-cells in the model are able to learn up to 150 different patterns, suggesting that the cerebellum can learn an enormous number of patterns in this way.

These ideas have been supported by experiments. LTD does shorten the pauses (Steuber *et al.* 2007). In order for the model to simulate this result, it was necessary for the parallel-fiber inputs to be synchronously active. They can be, thanks to the inhibitory interneurons.

We already know that the synaptic weights of the inputs to stellate and basket cells can be modified by learning. Stellate and basket cells are good timekeepers. They respond very quickly to their synaptic inputs, with changes in firing (Jorntell *et al.* 2010). They also inhibit each other, which keeps their responses short. Furthermore, interneurons can receive input from the same climbing fibers, and they are sometimes directly coupled by gap junctions (Mann-Metzer & Yarom 1999), which promotes synchronous firing. This means that molecular-layer interneurons are useful for shaping the timing of P-cell responses, particularly the timing of the pauses in firing that P-cells exhibit. Indeed, because basket cells innervate many P-cells (see Figure 1.1), a row of P-cells that runs in the parasagittal plane (within a microzone) may pause synchronously.

Synaptic plasticity at the mossy fiber-to-granule cell synapse (within the "glomerulus") has also been described. In the "window-matching"

---

[6] If enough calcium-sensitive potassium channels are open, the plasma membrane will be hyperpolarized and the P-cell will not fire. If only a few channels are open, firing rate is limited. These channels are important regulators of firing in most, if not all, cerebellar neurons.

theory, recently proposed by de Zeeuw's group, the granule cell layer converts continuous mossy-fiber inputs into discrete bursts of action potentials by the parallel fibers. In this way, groups of mossy-fiber inputs that are part of different contexts can be segregated from each other, according to when they fire their bursts. Because Golgi-cell inhibition is expected to cut short the burst at some point, it seems ideal to match the burst to the interval before inhibition arrives; this is window-matching. In fact, plasticity at the mossy-fiber inputs to granule cells is bidirectional, and can control the timing of the granule-cell response to its input burst (Nieus *et al.* 2006). As a result, the bursts may be almost synchronous when they arrive at the P-cells. This will make it easier for the P-cell to learn input patterns (D'Angelo & De Zeeuw 2009).

## CONCLUSIONS

Does the cerebellum operate as a linear system or as a discrete timing system? The answer is that it does both. While the cerebellum does these things, it is also learning, identifying patterns, and recoding information. These processes are not mutually exclusive, but our names for various tasks lead to the illusion of an either-or situation. It may make more sense to think of the cerebellum as a large bank of computers, each one doing whatever tasks it is assigned by its input data. Synaptic plasticity gives it the capacity to learn accurately-timed responses to particular patterns of input activity. It also can optimize an adaptive filter (actually, many filters) that determine how time-varying signals are transmitted by the cerebellum, and a pattern-recognition device (or devices) that encode patterns as pauses in P-cell discharge. Using all of its tools, the cerebellum can learn to recognize situations, and to generate optimized behavioral responses to them. Once it has learned, the cerebellum can react much more quickly than the cerebral cortex can.

## REFERENCES

Albus, J.S. (1971) A theory of cerebellar function. *Math. Biosci.*, **10**, 25–61.

Braitenberg, V. (1961) Functional interpretation of cerebellar histology. *Nature*, **190**, 539–540.

Carpenter, R.H.S. (1972) Cerebellectomy and the transfer function of the vestibulo-ocular reflex in the decerebrate cat. *Proc. R. Soc. Lond.*, **181**, 353–374.

D'Angelo, E. & De Zeeuw, C.I. (2009) Timing and plasticity in the cerebellum: focus on the granular layer. *Trends Neurosci.*, **32**, 30–40.

Dean, P., Porrill, J. & Stone, J.V. (2002) Decorrelation control by the cerebellum achieves oculomotor plant compensation in simulated vestibulo-ocular reflex. *Proc. R. Soc. Lond.*, **269**, 1895–1904.

Dean, P., Porrill, J., Ekerot, C.-F. & Jorntell, H. (2010) The cerebellar microcircuit as an adaptive filter: experimental and computational evidence. *Nat. Rev. Neurosci.*, **11**, 30–43.

Eccles, J.C. (1977) An instruction-selection theory of learning in the cerebellar cortex. *Brain Res.*, **127**, 327–352.

Fujita, M. (1982) Simulation of adaptive modification of the vestibulo-ocular reflex with an adaptive filter model of the cerebellum. *Biol. Cybern.*, **45**, 207–214.

Granit, R. & Phillips, C.G. (1956) Excitatory and inhibitory processes acting upon individual Purkinje cells of the cerebellum in cats. *J. Physiol.*, **133**, 520–547.

Ito, M. (1972) Neural design of the cerebellar motor control system. *Brain Res.*, **40**, 81–84.

Jorntell, H. & Ekerot, C.-F. (2002) Reciprocal bidirectional plasticity of parallel fiber receptive fields in cerebellar Purkinje cells and their afferent interneurons. *Neuron*, **34**, 797–806.

Jorntell, H., Bengtsson, F., Schonewille, M. & De Zeeuw, C.I. (2010) Cerebellar molecular layer interneurons—computational properties and roles in learning. *Trends Neurosci.*, **33**, 524–532.

Kenyon, G.T. (1997) A model of long-term memory storage in the cerebellar cortex: a possible role for plasticity at parallel fiber synapses onto stellate/basket interneurons. *Proc. Natl. Acad. Sci. USA*, **94**, 14200–14205.

Kim, C.-H., Oh, S.-H., Lee, J.H., Chang, S.O., Kim, J. & Kim, S.J. (2012) Lobule-specific membrane excitability of cerebellar Purkinje cells. *J. Physiol.*, **590**, 273–288.

Lang, E.J., Sugihara, L., Welsh, J.P. & Llinas, R. (1999) Patterns of spontaneous Purkinje cell complex spike activity in the awake rat. *J. Neurosci.*, **19**, 2728–2739.

Lisberger, S.G. (1994) Neural basis for motor learning in the vestibulo-ocular reflex of primates: III. Computational and behavioral analysis of the sites of learning. *J. Neurophysiol.*, **72**, 974–998.

Llinas, R. & Sasaki, M. (1989) The functional organization of the olivocerebellar system as examined by multiple Purkinje cell recordings. *Eur. J. Neurosci.*, **1**, 587–602.

Mann-Metzer, P. & Yarom, Y. (1999) Electrotonic coupling interacts with intrinsic properties to generate synchronized activity in cerebellar networks of inhibitory interneurons. *J. Neurosci.*, **19**, 3298–3306.

Marr, D. (1969) A theory of cerebellar cortex. *J. Physiol.*, **202**, 437–470.

Mauk, M.D. & Donegan, N.H. (1997) A model of Pavlovian eyelid conditioning based on the synaptic organization of the cerebellum. *Learn. Mem.*, **4**, 130–158.

Medina, J.F. & Mauk, M.D. (1999) Simulations of cerebellar motor learning: computational analysis of plasticity at the mossy fiber to deep nucleus synapse. *J. Neurosci.*, **19**, 7140–7151.

Medina, J.F. & Mauk, M.D. (2000) Computer simulation of cerebellar information processing. *Nat. Neurosci.*, **3**, 1205–1211.

Miles, F.A. & Eighmy, B.B. (1980) Long-term adaptive changes in primate vestibuloocular reflex. I. Behavioral observations. *J. Neurophysiol.*, **43**, 1406–1425.

Mugnaini, E. (1983) The length of cerebellar parallel fibers in chicken and rhesus monkey. *J. Comp. Neurol.*, **220**, 7–15.

Nieus, T., Sola, E., Mapelli, J., Saftenku, E., Rossi, P. & D'Angelo, E. (2006) LTP regulates burst initiation and frequency at mossy fiber-granule cell synapses of rat cerebellum: experimental observations and theoretical predictions. *J. Neurophysiol.*, **95**, 686–699.

Pellionisz, A. & Llinas, R. (1979) Brain modeling by tensor network theory and computer simulation. The cerebellum: distributed processor for predictive coordination. *Neuroscience*, **4**, 323–348.

Robinson, D.A. (1976) Adaptive gain control of vestibuloocular reflex by the cerebellum. *J. Neurophysiol.*, **39**, 954–969.

Sakurai, M. (1987) Synaptic modification of parallel fibre-Purkinje cell transmission in *in vitro* guinea-pig cerebellar slices. *J. Physiol.*, **394**, 463–480.

Salin, P.A., Malenka, R.C. & Nicoll, R.A. (1996) Cyclic AMP mediates a presynaptic form of LTP at cerebellar parallel fiber synapses. *Neuron*, **16**, 797–803.

Sejnowski, T.J. (1977) Storing covariance with nonlinearly interacting neurons. *J. Math. Biol.*, **4**, 303–321.

Steuber, V., Mittman, W., Hoebeek, F.E., Silver, R.A., De Zeeuw, C.I., Hausser, M. & De Schutter, E. (2007) Cerebellar LTD and pattern recognition by Purkinje cells. *Neuron*, **54**, 121–136.

Welsh, J.P., Lang, E.J., Sugihara, I. & Llinas, R. (1995) Dynamic organization of motor control within the olivocerebellar system. *Nature*, **374**, 453–457.

## FURTHER READING

Marr, D. (1969) A theory of cerebellar cortex. *J. Physiol.*, **202**, 437–470.

Albus, J.S. (1971) A theory of cerebellar function. *Math. Biosci.*, **10**, 25–61.

Medina, J.F. & Mauk, M.D. (2000) Computer simulation of cerebellar information processing. *Nature Neurosci.*, **3**, 1205–1211.

D'Angelo, E. & De Zeeuw, C. I. (2009) Timing and plasticity in the cerebellum: focus on the granular layer. T. I. N. *S.*, **32**, 30–40.

Dean, P., Porrill, J., Ekerot, C.-F. & Jorntell, H. (2010) The cerebellar microcircuit as an adaptive filter: Experimental and computational evidence. *Nat. Rev. Neurosci.*, **11**, 30–43.

# 3 Plasticity in the Cerebellar Cortex

As Marr and Albus proposed, the cerebellum can learn because it is a network whose synaptic weights, and other characteristics, are plastic. The different forms of plasticity are the tools that the cerebellum has at its disposal in its task of improving performance. One in particular, cerebellar LTD, is understood in great depth. Others are only beginning to be understood. Chapters 3 and 4 will describe how plasticity works at the cellular level, and how the different mechanisms may interact to cause learning.

## CEREBELLAR LONG-TERM DEPRESSION

James Albus argued that the synapses between parallel fibers and Purkinje cells (or PF–PC synapses) should become depressed during motor learning. Subsequently, Masaki Sakurai showed that if both the parallel fibers and the climbing fibers were stimulated, the excitatory postsynaptic potential (EPSP) resulting from parallel-fiber stimulation became depressed for at least 50 minutes (Sakurai 1987). Masanobu Kano showed that depression occurred only at the synapses that had been active during the pairing protocol, while other synapses on the same Purkinje cell (P-cell) were unaffected (Kano & Kato 1987). This "synapse specificity" was a key feature of the Marr and Albus models.

The occurrence of LTD with stimulation of climbing and parallel fibers is actually quite surprising. In the cerebral cortex, synaptic plasticity seems to follow a rule which was suggested in 1949 by Donald Hebb:

> When an axon of cell A is near enough to excite cell B and repeatedly or persistently takes part in firing it, some growth process or metabolic change takes place in one or both cells such that A's efficiency as one of the cells firing B, is increased (Hebb 1949).

Hebb's postulate does not seem to operate in the cerebellar cortex. An action potential in the climbing fiber reliably results in a complex spike in the P-cell, and this results in a decreased, not increased, "efficiency" of

---

*The Cerebellum: Learning Movement, Language, and Social Skills*, First Edition. Dianne M. Broussard.

cell A (the granule cell) in firing cell B (the P-cell). This change is useful, if in fact the complex spike represents an error.

Of course, the existence of a cellular mechanism, no matter how perfect it seems, does not mean that it is used for learning (or for anything in particular). It is questionable whether LTD at the PF–PC synapse is, in fact, an important mechanism for cerebellar learning. On the other hand, there is some hope that the mechanisms underlying learning at the PF–PC synapse may apply at most other cerebellar synapses.

## THE CALCIUM TRIGGER

How does the P-cell detect that its climbing and parallel fibers are being activated at the same time, and how do the appropriate synapses become depressed? These questions have kept neurophysiologists happily occupied for many years. Early on, it was found that LTD could be brought about by the binding of glutamate to particular receptors during the complex spike. If both AMPA-type ionotropic glutamate receptors[1] and metabotropic glutamate receptors (mGluRs)[2] on the P-cell are activated in the presence of complex-spike activity, LTD will occur (Kano & Kato 1988). Depolarization of the P-cell by passing current directly into it can substitute for climbing-fiber activity (Linden *et al.* 1991).

The coincident climbing-fiber and parallel-fiber inputs result in changes in intracellular calcium in the P-cell, as outlined in Figure 3.1. For one thing, during the complex spike, calcium enters the P-cell through voltage-gated calcium channels. In addition, calcium is released from internal stores, as follows: when glutamate is bound, type 1 metabotropic glutamate receptor (mGluR1) activation, by way of a G-protein,[3] causes increased activity of phospholipase C in the P-cell, generating inositol triphosphate. This in turn causes calcium to be released from the internal stores (Figure 3.1). Calcium release and calcium entry are both required, but the calcium from both does not simply add together. Instead, when parallel- and climbing-fiber activation coincide, the total increase in calcium concentration is considerably *greater* than would be expected by

[1]Alpha-amino-3-hydroxyl-5-methyl-isoxazole-propanoic acid (AMPA) receptors contain a cation channel that is mostly permeable to sodium. It has a high conductance and its current has a short time course.

[2]Metabotropic receptors act via G-proteins and have a longer time course of activation, and also access more different types of intracellular pathways, than the ionotropic receptors.

[3]G-proteins are intracellular proteins that can bind to either guanosine diphosphate (GDP) or guanosine triphosphate (GTP). When they are activated, GDP is exchanged for GTP, and the G-protein initiates a signaling cascade.

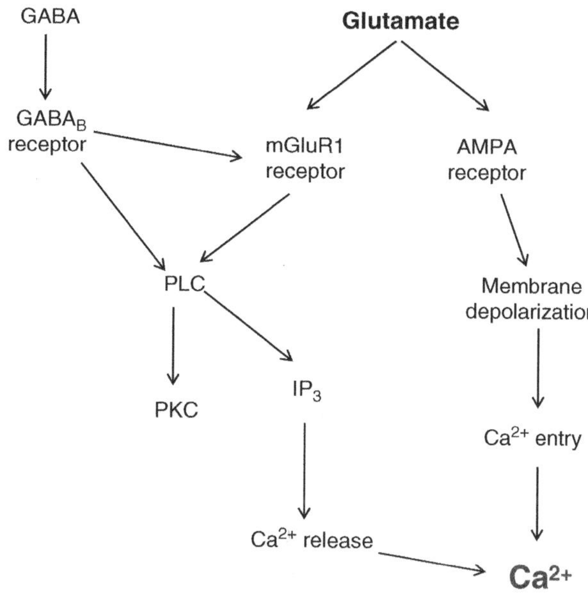


**Figure 3.1** Intracellular pathways affecting calcium concentration in dendritic spines of P-cells. Arrows indicate either a direct or an indirect effect, such as activation or increased concentration. See text for details.

adding them together (Wang *et al.* 2000), probably because of interactions between the complex spike and mGluR activation (Batchelor & Garthwaite 1997). Perhaps calcium is the trigger for LTD. If this is true, then an adequate calcium concentration in the cytosol reached by any means should cause LTD, and there is some evidence that this is the case (Lev-Ram *et al.* 1997).

To summarize, it appears that in the cerebellum, as in other brain regions, calcium is the key signal indicating that plasticity should take place. An important difference in the cerebellum is that instead of triggering potentiation, a high concentration of calcium in the cytosol of P-cells triggers LTD. The "calcium model" provides a built-in mechanism for detecting conjunctive activity of climbing and parallel fibers. Although parallel-fiber activity *can* activate all of the necessary receptor types, it is difficult to obtain LTD by stimulating only the parallel fibers. The very large depolarization associated with climbing-fiber input can boost the calcium concentration at the crucial moment. A similar argument may explain the requirement for activation of certain receptor types: under normal physiological conditions the receptors may all need to contribute, by way of their different mechanisms, to get the calcium concentration high enough.

Calcium is the trigger for LTD, but the process that it triggers is quite complex. Even though calcium influx probably lasts only a few seconds,

LTD can take up to an hour to develop (Konnerth *et al.* 1992). This slowness reflects the fact that several signaling cascades are involved. Signaling cascades are sequences of enzymatic reactions that serve to amplify responses, while also introducing points where the response can be controlled. Cascades that contribute to calcium concentration directly are shown in Figure 3.1. Other cascades reach across the synapse, and also affect calcium. Still more cascades are triggered by calcium and act on the postsynaptic glutamate receptors, inducing LTD.

In general, intracellular signaling pathways are largely controlled by protein kinases (which add phosphate groups to proteins) and phosphatases (which remove them). Several protein kinases are part of the LTD machine (Figures 3.1 and 3.3). Protein kinase C (PKC) (Linden & Connor 1991) and calcium/calmodulin-dependent kinase II (CaMKII) are two prominent examples (Hansel *et al.* 2006). Under the right circumstances, each of these kinases can contribute to LTD.

Let's take a moment to think about where in the P-cell these events are occurring. Glutamatergic synapses are located on tiny spines on the dendrites of P-cells. These spines partially segregate their own small volumes of cytosol from the rest of the dendrite. Spines also restrict electrical access to their own small regions of plasma membrane. mGluR1 and its associated cascades are located on and within spines, and so are the voltage-gated calcium channels. Although the segregation of the spine is not complete, it may be a factor in LTD. It may be the calcium concentration within each spine that determines whether LTD will occur in that spine (Wang *et al.* 2000). Increased calcium can even potentiate the $IP_3$ receptor's response, resulting in even more calcium release in a positive feedback mechanism, all within the spine.

One might think that completely specific, spine-based learning, based on these local increases in calcium, would be optimal. However, it turns out that LTD can spread to neighboring synapses, presumably located on other spines. Cerebellar LTD does not even occur at isolated synapses; instead, there must be interaction among synapses in a group. This is accomplished by more widely diffusible messengers, at least two of which cross the synaptic cleft. LTD requires some back-and-forth communication between neurons, analogous to a conversation. This conversation is mediated by at least four substances in the extracellular space: glutamate, gamma-amino-butyric acid (GABA), one or more endocannabinoids, and the gas nitric oxide (NO).

## THE SYNAPTIC CONVERSATION

After glutamate has crossed the synaptic cleft and triggered a rise in postsynaptic calcium, signals travel across the cleft two more times

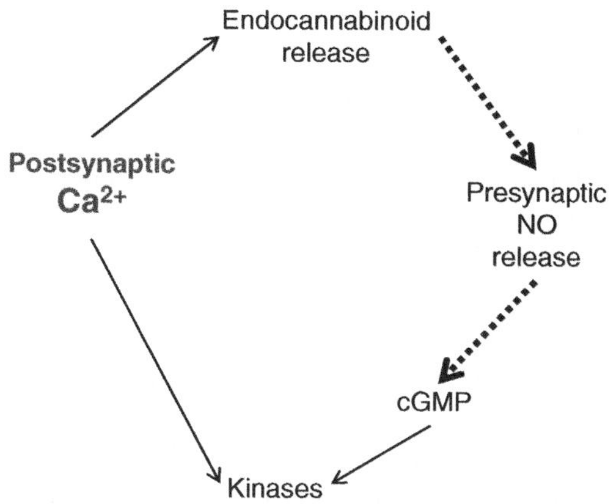


**Figure 3.2** Increased calcium concentration in the P-cell can activate kinases both directly and indirectly. The indirect pathway crosses the synaptic cleft twice (dotted arrows), first in the retrograde and then in the anterograde direction. See text for details.

(Figure 3.2). The cannabinoid receptor CB1 is located on the presynaptic terminals of parallel fibers. In order for LTD to occur, endocannabinoids must be released from the dendrites of P-cells (Figure 3.2) and diffuse "backward" across the synapse, as a retrograde messenger (Safo & Regehr 2005). Their release is triggered by activation of mGluR1 and elevation of calcium inside the P-cell. Next, if LTD is going to be expressed postsynaptically, an anterograde messenger must carry the message across the synapse again, to the P-cell dendrite. Action potentials in the parallel fibers cause an increase in the second messenger NO, a gas that diffuses easily in neural tissue. NO release is necessary for LTD (Shibuki & Okada 1991), and can activate protein kinases in the P-cell, by way of soluble guanylate cyclase and cyclic guanosine monophosphate (cGMP). In fact, calcium and NO can induce LTD without any electrical activity at all (Lev-Ram *et al.* 1997).

The induction of LTD by a diffusible gas would seem to explain the lack of synapse specificity in LTD. Instead of changing one synapse at a time, NO could allow recruitment, into a "depressed group", of additional synapses near those that have been subjected to pairing. With very strong parallel-fiber stimuli, NO does induce LTD at nearby synapses. Furthermore, LTD at neighboring synapses can be induced even without climbing fiber activation, so long as NO is released (Reynolds & Hartell 2000). Following a complex spike and large calcium influx into the P-cell, there is a corresponding potassium efflux and that increase in extracellular potassium can trigger LTD, which is blocked by inhibitors of NO synthase (Daniel *et al.* 1998).

So far I have described two ways of obtaining LTD, one that is synapse-specific and requires high postsynaptic calcium, and one that spreads to neighboring synapses (up to 100 μm away) and is mediated by NO. Although it is easier to imagine these two mechanisms as separate alternatives, they may interact. Isolated synapses, like those made by granule-cell axons as they ascend radially in the molecular layer (and before they bifurcate) do not seem to be capable of LTD (Marcaggi & Attwell 2007).

The stellate and basket cells also play a part in cerebellar cortical plasticity. In fact, they participate in multiple ways. To begin with, N-methyl-D-aspartic acid (NMDA) receptors make a contribution to LTD, even though they are not located at the modified synapses (Shin & Linden 2005); instead, they are probably located on the inhibitory interneurons. Interneurons may also participate in the triggering of cascades within the P-cell dendritic spines, by releasing GABA and activating $GABA_B$ receptors (Figure 3.1). Activation of $GABA_B$ receptors,[4] located on spines, enhances LTD by at least two pathways: a direct effect on the electrical component of the mGluR1 response, and an indirect effect via a signalling cascade that augments calcium release from intracellular stores (Kamikubo *et al.* 2007).

## THE MEMORY TRACE

LTD requires a complex sequence of molecular signals. Intracellular messengers (such as PKC) can remain elevated for tens of minutes (Tanaka & Augustine 2008). During the time that PKC activity remains high, the excitatory postsynaptic current (EPSC) that is triggered by glutamate becomes smaller (Linden *et al.* 1991). In the end, LTD is preserved as changes in glutamate-receptor function, specifically the function of the AMPA receptor.

Although LTD that is induced by 1-Hz stimulation is always expressed postsynaptically, presynaptic LTD can be induced by stimulation at other frequencies. Endocannabinoids also can cause a variant of LTD that affects the probability of transmitter release by the presynaptic terminal (Qiu & Knopfel 2009). Presynaptic LTD has been little studied to date, but it appears to require higher-frequency stimulation of parallel fibers, activation of PKA, and activation of NMDA receptors, that cooperatively trigger the release of endocannabinoids.

Phosphorylation of the AMPA receptor is certainly necessary for postsynaptic LTD (Chung *et al.* 2003) (Figure 3.3). PKC and PKA both can

[4]$GABA_B$ receptors are GABA receptors that do not contain ion channels, but work by way of second messengers.

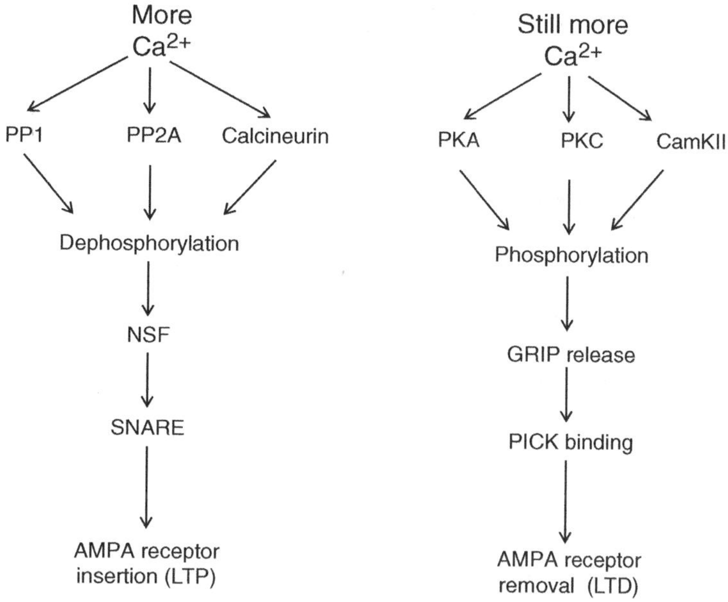


**Figure 3.3** Regulation of the number of AMPA receptors by kinases and phosphatases. Both kinase and phosphatase cascades are triggered by calcium, but the threshold is lower for the phosphatase pathway. The kinase pathway leads to LTD, and the phosphatase pathway leads to LTP. PK, protein kinase; PP, protein phosphatase; NSF, SNARE, GRIP, and PICK are proteins. See text for details.

phosphorylate the receptor. During LTD, there is a decrease in the number of AMPA receptors underlying the synapse (Wang & Linden 2000). This is because phosphorylation leads to removal of receptor molecules from the synapse.

Removal of AMPA receptors is mediated by the signalling cascade shown in Figure 3.3. When the AMPA receptor is phosphorylated, it detaches from an associated protein known as glutamate-receptor-interacting protein (GRIP). Once this happens, the AMPA receptors can be removed from the plasma membrane by endocytosis (Chung *et al.* 2003). The PICK1 protein is involved in the phosphorylation of the receptor by PKC (Steinberg *et al.* 2006) and also promotes the actual endocytosis (Leitges *et al.* 2004). Although endocytosis is continually going on, it is balanced by the delivery of new pieces of plasma membrane that may contain more, or fewer, AMPA receptors.

Whenever this membrane-cycling process becomes unbalanced, the synapse changes its strength. The balance can be shifted toward endocytosis by reducing the activity of the relevant phosphatases, and inhibition of phosphatases for which the AMPA receptor is a substrate can induce LTD (Ito 2001). For example, inhibition of the phosphorylase PP2A induces LTD (Launey *et al.* 2004). This may occur because more AMPA receptors are phosphorylated, or because kinases are inactivated by PP2A, or both.

## WHAT ABOUT POTENTIATION?

As I mentioned in Chapter 2, LTD should be reversible if we are to maximize the capacity of the network. In fact, as Figure 3.3 illustrates, the parallel fiber-to-P-cell synapse can be potentiated as well as depressed. If the parallel fibers fire but the climbing fibers do not, LTP occurs instead of LTD (Sakurai 1987). Cerebellar LTP can be either pre- or postsynaptic. Postsynaptic LTP requires prolonged stimulation of the parallel fibers at 1 Hz (Lev-Ram *et al.* 2002). Chelation of calcium in the P-cell also helps to potentiate the synapse, and in fact a smaller calcium increase seems to be required for LTP than for LTD (Coesmans *et al.* 2004). If calcium does not reach the threshold for LTP, then no change will occur. This is the inverse of the "BCM rule," which applies at hippocampal synapses, and which was originally proposed by Bienenstock *et al.* (1982); the BCM rule sets a higher calcium threshold for LTP than for LTD.

If the calcium threshold for LTD is to be reached, complex spikes must roughly coincide with parallel-fiber inputs, opening the dendritic voltage-gated calcium channels. LTP, on the other hand, is induced by low-frequency parallel-fiber stimulation alone. However, postsynaptic LTP and postsynaptic LTD have some mechanisms in common. Both require mGluR1 activation (Wang *et al.* 2009) as well as NO release (Lev-Ram *et al.* 2002). LTD and LTP diverge in the last segment of the signalling pathway (Figure 3.3). Instead of kinases, phosphatases are required for LTP, specifically PP1, PP2A, and calcineurin (PP2B) (Belmeguenai & Hansel 2005).

There are two isoforms of calcium/calmodulin-dependent kinase type II, $\alpha$CamKII and $\beta$CamKII. $\beta$CamKII predominates in the cerebellum. During LTP, the calcium concentration fails to reach the threshold for $\alpha$CamKII activation (Hansel *et al.* 2006), but $\beta$CamKII can determine the direction of plasticity under some circumstances, and phosphatase activity is also important (van Woerden *et al.* 2009). In general we can imagine that the direction of parallel-fiber plasticity is under the control of a "kinase-phosphatase switch," in which the relative activity levels of these two groups of enzymes, by controlling the proportion of phosphorylated GluR2 residues, determine whether AMPA receptors are preferentially trafficked into, or out of, synapses.

In summary, low calcium concentration in the vicinity of the synapse (possibly within the spine) means that no plasticity will occur. More calcium enables LTP, and more still, LTD. If the LTP threshold is not reached, no learning occurs.

The different processes that are triggered for LTD and LTP involve several of the proteins that carry out receptor trafficking. The adaptor protein PICK1 is involved in LTD, as already mentioned. In postsynaptic LTP, the relevant adaptor protein is NSF.[5] NSF attaches to SNARE, allowing the exocytosis of AMPA receptors (Kakegawa & Yuzaki 2005). In fact, NSF is required for maintaining basal synaptic transmission, because AMPA receptors are continually turning over.

In addition to the balance of exocytosis and endocytosis, there is another type of balance to consider: the balance of receptor trafficking within the plasma membrane. Receptors can be moved quickly into and out of the postsynaptic density. Trafficking between the postsynaptic density and an extrasynaptic, but membrane-bound, pool of receptor molecules may also contribute to synaptic plasticity.

If LTP is to increase the storage capacity of the network, it must completely reverse LTD. There is reason to think that it may. In fact, postsynaptic LTD and LTP may act on the same receptor molecules. The memory trace for LTP, like that for LTD, is constructed by trafficking of AMPA receptors that contain GluR2 (Kakegawa & Yuzaki 2005). Of course, reversal under some circumstances does not mean that it occurs under all circumstances. The fact that GluR2 subunits are involved does not mean that the same molecules are involved. If the PF–PC synapse is capable of truly reversible long-term synaptic plasticity, this makes it even more attractive as a locus of cerebellar motor learning. This important issue is still not fully resolved.

## *Spike patterns and memory location*

Cerebellar synaptic plasticity is associative, in the sense that it links different types of information. These different types of information often converge at the P-cell. In order to meet the calcium threshold for LTD or even for LTP, these convergent inputs must co-vary to some extent. That is, the discharge patterns of the input neurons, which may include parallel fibers and climbing fibers, must be correlated. This means that when one input is increasing, the other one should be increasing as well. In fact, in order for LTD to occur, the presynaptic neuron (the granule cell) must fire at around the same time that the P-cell fires a complex spike. This could be accomplished using 1-Hz stimulation, with climbing fibers stimulated 10 ms before the parallel fiber (Sakurai 1987), or using conjunctive stimulation at higher frequencies, or even burst stimulation of parallel fibers

---

[5]NSF stands for N-ethylmaleamide-sensitive factor. SNARE stands for NSF attachment protein receptor. This same complex is also involved in release of transmitter from the presynaptic terminal (however, it does not seem to be involved in presynaptic LTP).

paired with a pair of stimuli applied to the climbing fiber pathway (Wang *et al.* 2000). Of these different protocols, the burst stimulation is the most effective, but all of them work. In this respect, cerebellar LTD appears to be fairly forgiving. This is because the calcium signals that are required for LTD have quite slow dynamics (Tanaka *et al.* 2007), which effectively extend the time window for plasticity to occur.

Even though LTD is forgiving, it does have a preference. It turns out that the size of the increase in calcium concentration depends on spike timing. For the best results, the parallel fibers should be activated 50–200 ms before the climbing fiber. This relative timing maximizes both the calcium transient and the degree of LTD that is induced (Wang *et al.* 2000). The cellular component that enforces the timing rule is the inositol triphosphate receptor, which is inactivated by high levels of calcium that *precede* binding of $IP_3$. This means that in order for parallel fiber and climbing fiber coincidence to be detected (as in Figure 3.1), the mGluR1 receptor ideally should be activated before the voltage-gated calcium channels open. Since the climbing fiber input causes the opening of voltage-gated channels, this also means that the parallel-fiber inputs should be active before the climbing fiber fires, in order to maximize LTD.

In addition to the postsynaptic memory trace on P-cells, which can be either potentiation or depression, the parallel-fiber terminal can maintain potentiation (Salin *et al.* 1996).[6] Presynaptic LTP is expressed as greater-than-normal calcium entry during the release of neurotransmitter vesicles (Qiu & Knopfel 2007). It is induced by parallel-fiber stimulation at 4–8 Hz. Like postsynaptic LTP, it does not require climbing-fiber activation.

Presynaptic LTP is triggered by a calcium increase in the parallel fiber that results in increased levels of cyclic AMP. NMDA receptors (on interneurons) play some part in presynaptic LTP, but what they do exactly has remained elusive so far. PKA, activated by cyclic adenosine monophosphate (cAMP), is also a key element in the cascade for LTP (Linden & Ahn 1999) and is thought to act in part by phosphorylating RIM1$\alpha$, an active-zone protein (Lonart *et al.* 2003). Presynaptic LTP is prevented by endocannabinoids, the retrograde messengers that trigger LTD (Figure 3.2) (van Beugen *et al.* 2006). This safety factor may act to prevent LTP that might otherwise attenuate the effects of postsynaptic LTD.

[6]The location of synaptic plasticity in these LTP studies was inferred by two methods: whether or not paired-pulse facilitation (PPF) was affected, and whether or not EPSC amplitude was altered. PPF is a presynaptic mechanism that depends on the probability of release of transmitter, given an action potential in the presynaptic axon. PPF is therefore affected by forms of plasticity that alter release probability. EPSC amplitude, on the other hand, is affected by changes in postsynaptic receptor density and/or sensitivity.

## OTHER SITES OF PLASTICITY

Although LTD at the PF–PC synapse has been a good model system for understanding mechanisms of plasticity in the cerebellum, it is far from clear that it is the only, or even the major, mechanism for cerebellar learning. For one thing, most of the other synapses in the cerebellar cortex exhibit some form of long-term plasticity, and it seems likely that some of these also contribute to learning.

Let's start with the climbing fiber. Over the years, cerebellar physiologists have become quite comfortable with the idea that the climbing fiber is a teacher, modifying the parallel-fiber inputs where learning occurs. However, the climbing-fiber synaptic weight can also change. In fact, it turns out that the climbing-fiber–P-cell synapse is capable of LTD that is strikingly similar to parallel-fiber LTD (Hansel & Linden 2000). LTD is induced by stimulation of climbing fibers at rates that are within the physiological range (1–10 Hz). It requires mGluR1 activation, PKC activation, and a large postsynaptic calcium increase. It is expressed postsynaptically as a change in AMPA receptors (Shen *et al.* 2002). Climbing-fiber LTD does not seem to spread to neighboring parallel-fiber inputs, perhaps because it does not involve NO release.

Because climbing-fiber LTD affects the transient calcium influx that occurs during the later components of the complex spike (Weber *et al.* 2003), it could prevent parallel-fiber LTD. If calcium channels failed to open during the latter part of the complex spike, LTP might replace LTD at the parallel-fiber inputs to the affected P-cell. If so, then climbing-fiber LTD is a form of metaplasticity (i.e., plasticity that affects, or controls, the likelihood of another type of plasticity).

Repeated activation of the climbing fibers alone has another effect: it causes long-term depression of the late component of the response to parallel-fiber stimulation, a component that is usually ascribed to activation of mGluR1.[7] The reduction in the mGluR1 response is accompanied by a reduction in calcium release from internal stores, and completely blocks subsequent parallel-fiber LTD (Jin *et al.* 2007).

To summarize, two forms of metaplasticity are brought about by repeated activation of the climbing fiber alone, and both make parallel-fiber LTD less likely to happen in the future.

### *Granule-cell plasticity*

As already mentioned, synapses on most, if not all neuronal types in the cerebellar cortex are capable of plasticity. Most physiologists focus on the

[7] Activation of mGluRs results in opening of cation channels by way of second messengers.

P-cell synapses and ignore the other synapses. Nevertheless, the synapses linking mossy fibers to granule cells, in the cerebellar glomeruli, exhibit long-term plasticity (D'Angelo *et al.* 1999). Stimulation of mossy fibers at different frequencies can either potentiate or depress glutamatergic transmission at the mossy fiber–granule cell synapse (D'Angelo & De Zeeuw 2008). This type of LTP is presynaptic and can be reversed by presynaptic LTD (Gao *et al.* 2012).

Unlike parallel-fiber LTP, granule-cell LTP requires the activation of postsynaptic NMDA receptors,[8] resulting in calcium influx, and also requires mGluR1 and PKC activation. Less granule-cell activation is required for LTD than for LTP, suggesting that this type of LTP follows the BCM rule. LTP is increased by acetylcholine (Gao *et al.* 2012). It is also reduced by GABA, suggesting that the direction of plasticity may be under the control of the Golgi cells (Figures 1.1 and 1.2). Both AMPA and NMDA components of excitatory synaptic transmission are affected. LTP and LTD also display an interesting spatial pattern in the granular layer, in which potentiated regions tend to be surrounded by depressed regions.

## INTERNEURONS

The cerebellar molecular layer contains a powerful inhibitory circuit. Stellate and basket cells, like P-cells, receive glutamatergic transmission from parallel fibers, and these interneurons also receive collaterals from climbing fibers (Figure 1.2) (Sugihara *et al.* 1999). Multiple synaptic terminals of the basket cell surround the soma of the P-cell and the initial segment of its axon (Figure 1.1). Changes in this feed-forward inhibitory pathway are likely to cause changes in downstream processing. When Albus included the inhibitory interneurons of the cerebellar cortex in his neural-network model, the power of the model was much improved.

The requirements for LTD and LTP at the excitatory synapse between the parallel fiber and the interneuron are the reverse of the requirements for PF–PC plasticity, and more like the rules at the mossy-fiber to granule-cell synapses. In stellate and basket cells, a larger increase in the cytosolic calcium concentration is required for LTP than for LTD. LTD is induced at the same synapses by a slower stimulation rate than is required for LTP. LTD is dependent on mGluR2 and mGluR3, which is unusual

[8]The N-methyl-D-aspartate receptor is a type of glutamate receptor that contains an ion channel that is permeable to calcium as well as sodium. The NMDA receptors are coincidence detectors controlling the induction of LTP at many synapses throughout the brain.

(Rancillac & Crepel 2003); it is also helped along by calcium entry through calcium-permeable AMPA receptors. To establish the memory trace, the subunit composition of the AMPA receptors may be altered (using GRIP and PICK1) to a form that is less calcium-permeable, which actually limits further LTD (Liu & Cull-Candy 2002). Parallel fiber–interneuron LTD is highly synapse-specific.

Postsynaptic depolarization of the interneuron while the parallel fibers are active can shift the balance of postsynaptic plasticity toward LTP (Rancillac & Crepel 2003). Postsynaptic LTP is dependent on NMDA receptors and on either cAMP or NO. But LTP can also be presynaptic, in which case depolarization is not needed.

Coincident parallel and climbing-fiber inputs to interneurons may cause LTP of those inputs, and parallel-fiber inputs alone, LTD. During parallel-fiber LTP and LTD in whole animals, interneurons show changes in their receptive fields that mirror those of P-cells. The LTD protocol (paired parallel-fiber and climbing-fiber stimulation) causes a decrease in the receptive-field sizes of P-cells and an increase in the receptive fields of inhibitory neurons. LTP protocols reverse both of these changes (Jorntell & Ekerot 2002).

The result of all this is interesting. Patterns of stimulation that are meant to mimic natural "baseline" discharge patterns induce reciprocal changes, with LTP at the P-cell and LTD at the interneuron. On the other hand, burst-firing patterns that mimic sensory input (such as whisker deflection) potentiate both the excitatory and inhibitory arms of the feed-forward circuit (Smith & Otis 2005). This results in excitatory and inhibitory responses that arrive at different times. What it all means is that changes in the feed-forward inhibitory and excitatory pathways will never simply "cancel" and have no net effect. Instead, they will change what happens to signals passing through the circuit.

Golgi cells are powerful inhibitory interneurons that cross the boundary between the molecular and granular layers, providing inhibitory feedback from the parallel fibers to the glomerulus. It turns out that the excitatory inputs to Golgi cells are modifiable by yet another mechanism with its own set of rules. It is postsynaptic, depends on group II mGluRs, and yet is independent of NMDA receptors. It may be guided by climbing-fiber inputs to Golgi cells (Gao *et al.* 2012).

### *Inhibitory synaptic plasticity*

Can modification of inhibitory transmission in the cerebellum contribute to learning? This possibility had not been seriously considered when Masanobu Kano discovered plasticity of the inhibitory synapses between inhibitory interneurons and P-cells (Kano *et al.* 1992). Kano found that following climbing-fiber stimulation at a low frequency (two per second), the amplitude of spontaneous IPSCs[9] in P-cells was nearly doubled, and this change persisted for at least 40 minutes. Kano called the long-term potentiation of inhibitory synapses "rebound potentiation," because it counteracted the usual decline in IPSC amplitude over the recording period. LTP of inhibitory synapses requires a transient increase in calcium in the P-cell dendrite, involving the release of calcium from internal stores. Kano's idea was that it might contribute to motor learning, and he was probably right.

LTP at the interneuron–P-cell synapse depends on the activation of CamKII by the calcium transient (Kano *et al.* 1996). PKA activity is also required (Kano 1996). Interestingly, the PKA pathway is suppressed by the presence of the inhibitory transmitter itself, suggesting that synapses that are inactive are more likely to become potentiated (Kawaguchi & Hirano 2000). Inhibitory LTP is stored as a change in sensitivity of the postsynaptic response to GABA, which involves other proteins and probably involves GABA-receptor trafficking.

The inhibitory synapses between interneurons and P-cells can be depressed as well as potentiated, although it remains unclear exactly what controls the direction of change. The same coincident firing of parallel fibers and climbing fibers that induces parallel-fiber LTD also induces LTD in the inhibitory feed-forward pathway onto the same P-cells (Mittman & Hausser 2007). This suggests that excitatory LTD may be balanced by inhibitory LTD. The outcome of all this may be a change in signaling rather than a shift in the average firing rate.

## *Intrinsic plasticity*

While synaptic plasticity has been the focus of most research on learning mechanisms, neurons can also change their responsiveness to input pathways by other means. For example, their excitability can change. Intrinsic excitability is the likelihood that input from an excitatory synapse or synapses (or injected current) will actually cause the cell to fire. It depends on factors like the electrical resistance of the plasma membrane and the threshold membrane potential for generating action potentials.

Intrinsic excitability is known to be modified in P-cells during the protocols that induce LTP (Belmeguenai *et al.* 2010). Calcium signaling and activation of phosphatases are necessary for this modification. LTP itself also results in increased excitability because it downregulates certain potassium channels (the SK channels) that regulate firing directly;

[9]IPSCs are inhibitory postsynaptic currents. These are recorded in voltage-clamp mode, with a patch electrode, usually from slice preparations. There is a constant barrage of IPSCs arriving at P-cells due to tonic activity of inhibitory interneurons.

LTP removes one of the cell's braking mechanisms. Because increased excitability causes the P-cell to be more depolarized it may prevent further LTP. The outcome of this combination of mechanisms may be increased synapse specificity for LTP. The membrane resistance and spike threshold of granule cells is modified during mossy-fiber LTP, which should amplify the changes (Gao *et al.* 2012).

Because intrinsic plasticity affects the whole postsynaptic cell, rather than individual synapses or groups of synapses, its value for learning is less obvious than the value of synaptic plasticity. One possibility is that intrinsic plasticity improves the signal-to-noise ratio and other properties of the P-cell discharge pattern. More research is needed in order to understand the mechanisms and importance of intrinsic plasticity in the cerebellar cortex.

## CONCLUSIONS

Between synaptic LTP and LTD at excitatory and inhibitory synapses and intrinsic plasticity, the cerebellar cortex has a large toolbox at its disposal for learning. As already mentioned, the cortical circuitry may carry out expansion recoding by generating a variety of temporal patterns in the granular layer. Using the distributed and synergistic plasticity (Gao *et al.* 2012) that is present at these various sites, it may also carry out context-based learned selection among these particular patterns. It also seems likely that the cerebellum can modify the filtering properties of its intrinsic network as necessary to improve the accuracy and timing of movements.

In Chapters 5 and 6, I will describe the actual evidence linking some of these mechanisms to learning. It is important to remember that with such a multitude of tools, if one of them is lost, the others may be able to compensate. It is possible that no single mechanism is essential, but that it is essential to have most of them.

It is possible that because changes are easy to induce in the cerebellar cortex, it is not a good site for retention of memory in the long term. Changes can also occur in the DCN, but they seem to be responsible for forms of learning that are particularly persistent. These forms of learning require a slightly different set of mechanisms, which will be described in the next chapter.

## REFERENCES

Batchelor, A.M. & Garthwaite, J. (1997) Frequency detection and temporally dispersed synaptic signal association through a metabotropic glutamate receptor pathway. *Nature*, **385**, 74–77.

Belmeguenai, A. & Hansel, C. (2005) A role for protein phosphatases 1, 2A, and 2B in cerebellar long-term potentiation. *J. Neurosci.*, **25**, 10768–10772.

Belmeguenai, A., Hosy, E., Bengtsson, F., Pedroarena, C.M., Piochon, C., Teuling, E., He, Q., Ohtsuki, G., De Jeu, M.T.G., Elgersma, Y., De Zeeuw, C.I., Jorntell, H. & Hansel, C. (2010) Intrinsic plasticity complements long-term potentiation in parallel fiber input gain control in cerebellar Purkinje cells. *J. Neurosci.*, **30**, 13630–13643.

Bienenstock, E.L., Cooper, L.N. & Munro, P.W. (1982) Theory for the development of neuron selectivity: orientation specificity and binocular interaction in visual cortex. *J. Neurosci.*, **2**, 32–48.

Chung, H.J., Steinberg, J.P., Huganir, R.L. & Linden, D.J. (2003) Requirement of AMPA receptor GluR2 phosphorylation for cerebellar long-term depression. *Science.*, **300**, 1751–1755.

Coesmans, M., Weber, J.T., de Zeeuw, C. & Hansel, C. (2004) Bidirectional parallel fiber plasticity in the cerebellum under climbing fiber control. *Neuron.*, **44**, 691–700.

D'Angelo, E., Rossi, P., Armano, S. & Taglietti, V. (1999) Evidence for NMDA and mGlu receptor-dependent long-term potentiation of mossy fiber-granule cell transmission in rat cerebellum. *J. Neurophysiol.*, **81**, 277–287.

D'Angelo, E. & De Zeeuw, C.I. (2008) Timing and plasticity in the cerebellum: focus on the granular layer. *Trends Neurosci.*, **32**, 30–40.

Daniel, H., Levenes, C. & Crepel, F. (1998) Cellular mechanisms of cerebellar LTD. *Trends Neurosci.*, **21**, 401–407.

Gao, Z., van Beugen, B.J. & De Zeeuw, C.I. (2012) Distributed synergistic plasticity and cerebellar learning. *Nat. Rev. Neurosci.*, **13**, 619–635.

Hansel, C. & Linden, D.J. (2000) Long-term depression of the cerebellar climbing fiber-Purkinje neuron synapse. *Neuron*, **26**, 473–482.

Hansel, C., de Jeu, M., Belmeguenai, A., Houtman, S.H., Buitendijk, G.H., Andreev, D., De Zeeuw, C. & Elgersma, Y. (2006) aCaMKII is essential for cerebellar LTD and motor learning. *Neuron*, **51**, 835–843.

Hebb, D.O. (1949) *The Organization of Behavior: A Neuropsychological Theory*. Wiley, New York.

Ito, M. (2001) Cerebellar long-term depression: characterization, signal transduction, and functional roles. *Physiol. Rev.*, **8**, 1143–1195.

Jin, Y., Kim, S.J., Kim, J., Worley, P.F. & Linden, D.J. (2007) Long-term depression of mGluR1 signaling. *Neuron*, **55**, 277–287.

Jorntell, H. & Ekerot, C.F. (2002) Reciprocal bidirectional plasticity of parallel fiber receptive fields in cerebellar Purkinje cells and their afferent interneurons. *Neuron*, **34**, 797–806.

Kakegawa, W. & Yuzaki, M. (2005) A mechanism underlying AMPA receptor trafficking during cerebellar long-term potentiation. *Proc. Nat. Acad. Sci. USA*, **102**, 17846–17851.

Kamikubo, Y., Tabata, T., Kakizawa, S., Kawakami, D., Watanabe, M., Ogura, A., Iino, M. & Kano, M. (2007) Postsynaptic $GABA_B$ receptor signalling enhances LTD in mouse cerebellar Purkinje cells. *J. Physiol.*, **585**, 549–563.

Kano, M. & Kato, M. (1987) Quisqualate receptors are specifically involved in cerebellar synaptic plasticity. *Nature*, **325**, 276–279.

Kano, M. & Kato, M. (1988) Mode of induction of long-term depression at parallel fibre-Purkinje cell synapses in rabbit cerebellar cortex. *Neurosci. Res.*, **5**, 544–556.

Kano, M., Rexhausen, U., Dreessen, J. & Konnerth, A. (1992) Synaptic excitation produces a long-lasting rebound potentiation of inhibitory synaptic signals in cerebellar Purkinje cells. *Nature*, **356**, 601–604.

Kano, M. (1996) Long-lasting potentiation of GABAergic inhibitory synaptic transmission in cerebellar Purkinje cells: its properties and possible mechanisms. *Behav. Brain Sci.*, **19**, 354–361.

Kano, M., Kano, M., Fukunaga, K. & Konnerth, A. (1996) Ca2+-induced rebound potentiation of g-aminobutyric acid-mediated currents requires activatigon of Ca2+/calmodulin-dependent kinase II. *Proc. Nat. Acad. Sci. USA*, **93**, 13351–13356.

Kawaguchi, S. & Hirano, T. (2000) Suppression of inhibitory synaptic potentiation by presynaptic activity through postsynaptic $GABA_B$ receptors in a Purkinje neuron. *Neuron*, **27**, 339–347.

Konnerth, A., Dreessen, J. & Augustine, G.J. (1992) Brief dendritic calcium signals initiate long-lasting synaptic depression in cerebellar Purkinje cells. *Proc. Nat. Acad. Sci. USA*, **89**, 7051–7055.

Launey, T., Endo, S., Sakai, R., Harano, J. & Ito, M. (2004) Protein phosphatase 2A inhibition induces cerebellar long-term depression and declustering of synaptic AMPA receptor. *Proc. Nat. Acad. Sci. USA*, **101**, 676–681.

Leitges, M., Kovac, J., Plomann, M. & Linden, D.J. (2004) A unique PDZ ligand in PKCa confers induction of cerebellar long-term synaptic depression. *J. Neurosci.*, **44**, 585–594.

Lev-Ram, V., Jiang, T., Wood, J., Lawrence, D.S. & Tsien, R.Y. (1997) Synergies and coincidence requirements between NO, cGMP, and $CA^{2+}$ in the induction of cerebellar long-term depression. *Neuron*, **18**, 1025–1038.

Lev-Ram, V., Wong, S.T., Storm, D.R. & Tsien, R.Y. (2002) A new form of cerebellar long-term potentiation is postsynaptic and depends on nitric oxide but not cAMP. *Proc. Nat. Acad. Sci. USA*, **99**, 8389–8393.

Linden, D.J. & Connor, J.A. (1991) Participation of postsynaptic PKC in cerebellar long-term depression in culture. *Science*, **254**, 1656–1658.

Linden, D.J., Dickinson, M.H., Smeyne, M. & Connor, J.A. (1991) A long-term depression of AMPA currents in cultured cerebellar Purkinje neurons. *Neuron*, **7**, 81–89.

Linden, D.J. & Ahn, S. (1999) Activation of presynaptic cAMP-dependent protein kinase is required for induction of cerebellar long-term potentiation. *J. Neurosci.*, **19**, 10221–10227.

Liu, S.J. & Cull-Candy, S.G. (2002) Activity-dependent change in AMPA receptor properties in cerebellar stellate cells. *J. Neurosci.*, **22**, 3881–3889.

Lonart, G., Schoch, S., Kaeser, P.S., Larkin, S.J., Sudhof, T.C. & Linden, D.J. (2003) Phosphorylation of RIM1a by PKA triggers presynaptic long-term potentiation at cerebellar parallel fiber synapses. *Cell*, **115**, 49–60.

Marcaggi, P. & Attwell, D. (2007) Short- and long-term depression of rat cerebellar parallel fibre synaptic transmission mediated by synaptic crosstalk. *J. Physiol.*, **578**, 545–550.

Mittman, W. & Hausser, M. (2007) Linking synaptic plasticity and spike output at excitatory and inhibitory synapses onto cerebellar Purkinje cells. *J. Neurosci.*, **27**, 5559–5570.
Qiu, D.L. & Knopfel, T. (2007) An NMDA receptor/nitric oxide cascade in presynaptic parallel fiber-Purkinje neuron long-term potentiation. *J. Neurosci.*, **27**, 3408–3415.
Qiu, D.L. & Knopfel, T. (2009) Presynaptically expressed long-term depression at cerebellar parallel fiber synapses. *Pflugers Arch.*, **457**, 865–875.
Rancillac, A. & Crepel, F. (2003) Synapses between parallel fibres and stellate cells express long-term changes in synaptic efficacy in rat cerebellum. *J. Physiol.*, **554**, 707–720.
Reynolds, T. & Hartell, N.A. (2000) An evaluation of the synapse specificity of long-term depression induced in rat cerebellar slices. *J. Physiol.*, **527**, 563–577.
Safo, P.K. & Regehr, W.G. (2005) Endocannabinoids control the induction of cerebellar LTD. *Neuron*, **48**, 647–659.
Sakurai, M. (1987) Synaptic modification of parallel fibre-Purkinje cell transmission in in vitro guinea-pig cerebellar slices. *J. Physiol.*, **394**, 463–480.
Salin, P.A., Malenka, R.C. & Nicoll, R.A. (1996) Cyclic AMP mediates a presynaptic form of LTP at cerebellar parallel fiber synapses. *Neuron*, **16**, 797–803.
Shen, Y., Hansel, C. & Linden, D.J. (2002) Glutamate release during LTD at cerebellar climbing fiber-Purkinje cell synapses. *Nat. Neurosci.*, **5**, 725–726.
Shibuki, K. & Okada, D. (1991) Endogenous nitric oxide release required for long-term synaptic depression in the cerebellum. *Nature*, **349**, 326–328.
Shin, J.H. & Linden, D.J. (2005) An NMDA receptor/nitric oxide cascade is involved in cerebellar LTD but is not localized to the parallel fiber terminal. *J. Neurophysiol.*, **94**, 4281–4289.
Smith, S.L. & Otis, T.S. (2005) Pattern-dependent, simultaneous plasticity differentially transforms the input-output relationship of a feedforward circuit. *Proc. Nat. Acad. Sci. USA*, **102**, 14902–14906.
Steinberg, J.P., Takamiya, K., Shen, Y., Xia, J., Rubio, M.E., Yu, S., Jin, W., Thomas, G.M., Linden, D.J. & Huganir, R.L. (2006) Targeted in vivo mutations of the AMPA receptor subunit GluR2 and its interacting protein PICK1 eliminate cerebellar long-term depression. *Neuron*, **49**, 845–860.
Sugihara, I., Wu, H.-S. & Shinoda, Y. (1999) Morphology of single olivocerebellar axons labeled with biotinylated dextran amine in the rat. *J. Comp. Neurol.*, **414**, 131–148.
Tanaka, K., Khiroug, L., Santamaria, F., Doi, T., Ogasawara, H., Ellis-Davies, G.C.R., Kawato, M. & Augustine, G.J. (2007) Ca2+ requirements for cerebellar long-term synaptic depression: Role for a postsynaptic leaky integrator. *Neuron*, **54**, 787–800.
Tanaka, K. & Augustine, G.J. (2008) A positive feedback signal transduction loop determines timing of cerebellar long-term depression. *Neuron*, **59**, 608–620.
van Beugen, B.J., Nagaraja, R.Y. & Hansel, C. (2006) Climbing fiber-evoked endocannabinoid signaling heterosynaptically suppresses presynaptic cerebellar long-term potentiation. *J. Neurosci.*, **26**, 8289–8294.

van Woerden, G.M., Hoebeek, F.E., Gao, Z., Nagaraja, R.Y., Hoogenraad, C.C., Kushner, S.A., Hansel, C., De Zeeuw, C.I. & Elgersma, Y. (2009) bCaMKII controls the direction of plasticity at parallel fiber-Purkinje cell synapses. *Nat. Neurosci.*, **12**, 823–825.

Wang, S.S.-H., Denk, W. & Hausser, M. (2000) Coincidence detection in single dendritic spines mediated by calcium release. *Nat. Neurosci.*, **3**, 1266–1273.

Wang, X., Chen, G., Gao, W. & Ebner, T. (2009) Long-term potentiation of the responses to parallel fiber stimulation in mouse cerebellar cortex *in vivo*. *Neuroscience*, **162**, 713–722.

Wang, Y.T. & Linden, D.J. (2000) Expression of cerebellar long-term depression requires postsynaptic clathrin-mediated endocytosis. *Neuron*, **25**, 635–647.

Weber, J.T., De Zeeuw, C., Linden, D.J. & Hansel, C. (2003) Long-term depression of climbing fiber-evoked calcium transients in Purkinje cell dendrites. *Proc. Nat. Acad. Sci. USA*, **100**, 2878–2883.

## FURTHER READING

Ito, M. (2001) Cerebellar long-term depression: characterization, signal transduction, and functional roles. *Physiol. Rev.*, **8**, 1143–1195.

Rose, C.R. & Konnerth, A. (2001) Stores not just for storage: intracellular calcium release and synaptic plasticity. *Neuron*, **31**, 519–522.

Hansel, C., Linden, D.J. & D'Angelo, E. (2001) Beyond parallel fiber LTD: the diversity of synaptic and non-synaptic plasticity in the cerebellum. *Nat. Neurosci.*, **4**, 467–475.

Jorntell, H. & Hansel, C. (2006) Synaptic memories upside down: bidirectional plasticity at cerebellar parallel fiber-Purkinje cell synapses. *Neuron.*, **52**, 227–238.

Jorntell, H., Bengtsson, F., Schonewille, M. & De Zeeuw, C.I. (2010) Cerebellar molecular layer interneurons–computational properties and roles in learning. *Trends Neurosci.*, **33**, 524–532.

Gao, Z., van Beugen, B.J. & De Zeeuw, C.I. (2012) Distributed synergistic plasticity and cerebellar learning. *Nat. Rev. Neurosci.*, **13**, 619–635.

# 4 Adjusting the Memory Trace

When memories are first formed, they are vulnerable. New memories can be disrupted by strong stimuli or by other learning, or they can simply fade. Or, memories can consolidate. As the word implies, consolidation makes memories less vulnerable to disruption or loss by other means. Long-term memories are consolidated, and relatively stable. But even a long-term memory can be extinguished if it was first encoded in response to associated events, and one of those events repeatedly occurs alone. The process of extinction may involve new learning that obscures the original memory. In that case, the original memory trace still exists, and if the associated events happen again later, the memory may form more quickly the second time around. This effect is known as savings.

The long-term memory trace is also known as the "engram." Originally, this term referred to a record, analogous to a phonograph record. Where is the record located? Does the memory stay where it is first formed, or does it move? Motor memory in particular is famously long lasting; we never forget how to ride a bicycle, and motor skills like those involved in playing musical instruments (for example) are often stored unused for years. Even if we think we have lost these skills, it is often easy to reacquire them. In this chapter, I will explore some possible factors that may make cerebellar motor memories especially durable.

It is possible that some cognitive skills, such as language learning and math computation, may follow the same rules as motor skills. In other words, if these memories become stored in the cerebellum through extended practice, they may become especially durable. With this possibility in mind, let's look at how consolidation works in the cerebellum.

## CONSOLIDATION MECHANISMS

How does the cerebellum consolidate memories? It could be done locally in the cerebellar cortex, by modifying synapses so that memories are retained. If so, protein synthesis in Purkinje cells (P-cells) might be involved. Alternatively, the long-term memory could be moved to another

---

*The Cerebellum: Learning Movement, Language, and Social Skills*, First Edition. Dianne M. Broussard.

location, freeing the cortical circuitry to perform other computations; this is the memory transfer hypothesis. We have proposed mechanisms for learning both in the cortex and in the DCN, and some scientists think that the DCN has a unique role as a site for long-term storage. These two hypotheses—local consolidation in the cortex and long-term storage in the DCN—have both received some attention, and it seems likely that both mechanisms do, in fact, contribute to consolidation.

First, consider local protein synthesis. LTD at the PF–PC synapse is expressed as a decrease in the postsynaptic response that lasts 30–60 minutes in slice preparations. Of course, this is not adequate for long-term storage. In cultured cells, if protein synthesis is allowed, LTD can be maintained without any decrement for at least 2 hours and possibly much longer. But if either transcription or translation is prevented, the postsynaptic response returns to baseline within an hour or so. The more permanent form is known as the "late phase" of LTD (Linden 1996). Late-phase LTD seems to be a good first step toward memory consolidation. We will refer to such local processes in the cerebellar cortex as *in situ* consolidation.

In P-cells that have experienced LTD, protein synthesis can be enabled by the activation of transcription factors, like cyclic AMP response element-binding (CREB) protein and serum response factor. These factors contribute to the late phase of LTD. Calcium/calmodulin-dependent kinase IV (CaMKIV), which is activated during LTD, phosphorylates the CREB protein (Ahn *et al.* 1999) and activates it. Meanwhile, serum response factor supports late LTD via the Arc promoter (Smith-Hicks *et al.* 2010). It is not known why protein synthesis is necessary for late LTD. One possibility is that it is needed for synaptic tagging, a mechanism that marks particular synapses for consolidation. Tagging has been described in the cerebral cortex (Redondo & Morris 2011), but not yet in the cerebellum. Protein synthesis could also enable structural changes that directly stabilize the change in receptor number.

As already mentioned, the memory could be transferred to another site during consolidation. During memory transfer, changes would be made at the second site, which could be "downstream" from the initial site of memory encoding. For example, cerebellar memories could be transferred from P-cells to the DCN and the vestibular nuclei.

The DCN have only a tiny fraction of the number of cells and synapses that are available in the cerebellar cortex, and none of the specialized structures that are thought to facilitate learning (such as parallel fibers). These limitations suggest that the DCN can encode only a small fraction of the information that can be encoded in the cortex. How, then, can it be useful to transfer the memory to these less spacious brain regions? One possibility is that the cortical machinery is needed to select what memories should be stored, but that the actual storage requires less space. Also, the transfer hypothesis assumes that once memories have been transferred to the deep nuclei they are difficult to disrupt, so the mechanism must be selective about what is transferred.

An alternative scenario is that plasticity mechanisms in the deep nuclei allow amplification of the memory "signal" as it is transmitted out of the cerebellum, so that fewer P-cells are required to encode it. In this scenario the memory is not transferred, but sites in the deep nuclei are added to the memory trace. This possibility does not require that the original memory trace is destroyed, but it does require some mechanism of memory storage at the new location.

Although their mechanisms are necessarily different, *in situ* protein synthesis in the cerebellar cortex and memory transfer to the deep nuclei are not mutually exclusive. It is not an either/or situation, and both may contribute to cerebellar learning.

## MEMORY TRANSFER AND SYNAPTIC PLASTICITY

How can a memory be transferred from the cerebellar cortex to the deep nuclei? One way is that a learned change in the temporal pattern of P-cell firing could cause synaptic plasticity in the connections of P-cells to the deep nuclei, shifting the memory downstream. This theory is testable by a modeling approach, in which we assume that there is a spike-timing-dependent mechanism of plasticity in the DCN that can respond to changes in the pattern of P-cell spikes. We can then ask, how good is the model at predicting experimental results *in vivo*? This was done in a recent study from the de Zeeuw lab (Wulff *et al.* 2009). Both the model and the experiments showed that a lack of inhibitory transmission in the cerebellar cortex affected P-cell discharge patterns very subtly. However, it had devastating effects on cerebellar memory consolidation. This study demonstrated that memory transfer relying on inhibitory cortical interneurons, P-cell activity patterns, and synaptic plasticity in the DCN is a likely mechanism for memory consolidation.

What kind of synaptic plasticity can respond to changes in the P-cell spiking patterns? There are several candidate mechanisms in the DCN; the first to be discovered was plasticity at the GABAergic synapse between the P-cell and the DCN neuron. Stimulating the P-cell axons can silence the DCN neurons. After being silenced, a DCN neuron may fire a burst of spikes (the rebound). The rebound discharge causes calcium entry through voltage-gated channels (Muri & Knopfel 1994). Repeating this protocol causes either LTD or LTP of the inhibitory synapse (Aizenman *et al.* 1998). NMDA-receptor channels also contribute to inhibitory plasticity (Ouardouz & Sastry 2000).

The key to determining whether the P-cell–DCN synapse will be potentiated or depressed is the activity of the DCN neuron at the crucial moment. If the DCN neuron fires, LTP will occur and if it does not, LTD will occur. Inhibitory synaptic plasticity could, technically, transfer the memory to the DCN neuron, but it would still depend on the same P-cell that originally encoded it to provide the presynaptic signal. This does not seem especially useful, if the goal of memory transfer is to free up P-cells for other computations. Instead, inhibitory LTD and LTP may serve a homeostatic function, regulating the resting discharge rates of DCN neurons so that they stay within their linear ranges.[1] For example, if the DCN neuron fires at a high rate, then inhibitory LTP can increase the efficacy of the inhibitory input that is constantly bombarding it, reducing its firing rate.

## MOSSY-FIBER COLLATERALS IN THE DCN

If inhibitory LTD and LTP are indeed homeostatic, then we need to look elsewhere for mechanisms of memory transfer. Fortunately, the excitatory synapses between the collaterals of mossy fibers and DCN neurons also are capable of long-term plasticity. The rules for such plasticity are non-Hebbian. If mossy fibers are stimulated at the same time that the DCN neuron is *hyperpolarized*, then the excitatory synapses that connect the mossy fibers to the DCN neurons will undergo LTP (Pugh & Raman 2006). Plasticity seems to be controlled by the inhibitory input from P-cells. This inhibitory input hyperpolarizes the DCN neuron, and the mossy-fiber input that is active becomes potentiated (Pugh & Raman 2008). This makes sense when we consider that DCN neurons are spontaneously active at fairly high firing rates. In the DCN, there is no complex spike to regulate plasticity. A plasticity mechanism triggered by coincident excitatory input and postsynaptic firing would be constantly triggered and useless for learning.

How does it work, then? It appears that some of the rebound currents that occur after hyperpolarizing a DCN neuron are carried by calcium entering the cell, and the calcium helps to drive excitatory potentiation (Pugh & Raman 2006). CaMKII and calcineurin (a calcium-dependent phosphatase) both are necessary for LTP, and both are triggered by calcium influx through NMDA receptors. Calcium influx probably happens a good deal of the time in these cells. However, in order for increased influx to be effective in activating these enzymes, the *tonic* calcium influx (through L-type channels) must first be reduced, by hyperpolarization. Once this has happened, CaMKII can be activated by rebound firing, and calcineurin can be activated by mossy-fiber inputs. In fact, three events must occur in order: inactivation of L-type channels, followed by

[1]In this context, the linear range is the range of firing rates that permit input signals to be transmitted without distortion.

calcineurin "priming", followed by CaMKII "triggering". All three will occur only if the excitatory and inhibitory input signals to the DCN neuron are correctly timed. LTP is linked, by its requirement for a sequence of intracellular events, to correct timing of the excitatory and inhibitory input signals (Person & Raman 2010).

LTD also occurs at the same synapses. LTD requires mossy-fiber activity during postsynaptic depolarization, rather than hyperpolarization. It is calcium-dependent, but does not depend on NMDA receptors. Instead, it requires activation of mGluR1 receptors (Zhang & Linden 2006). All of this means that bidirectional plasticity is available not only in the cerebellar cortex, but also in the DCN. The existence of bidirectional plasticity in the DCN is encouraging if we want to think in terms of re-encoding memory in the DCN that was initially encoded by bidirectional plasticity in the cortex.

In later chapters we will be looking at some evidence that memory really is transferred to the DCN for long-term storage, and that memory is at least sometimes transferred to the other targets of P-cells, the vestibular nuclei. LTP and LTD of excitatory synapses can happen in the vestibular nuclei; some of these excitatory synapses are from collaterals of mossy fibers. The rules are different in the vestibular nuclei. Unlike LTP in both P-cells and DCN neurons, LTP in vestibular premotor neurons requires higher intracellular calcium than LTD. LTP requires calcium entry by way of calcium-permeable AMPA receptors paired with hyperpolarization, while LTD requires calcium entry through NMDA receptors, paired with depolarization (McElvain *et al.* 2010).

Hidden in the transfer hypothesis of memory consolidation is an assumption that once a memory is stored in the DCN, it will somehow be more persistent and more resistant to change. It is possible that the exacting requirements for LTD and LTP in the DCN simply make it less likely that changes will be reversed later on. Other possibilities also exist. As we saw in the cerebellar cortex, protein synthesis is required for "late" long-term plasticity in the DCN (Zhang & Linden 2006). One function of protein synthesis may be to physically change the synapses, either in size or number. The average size of excitatory synapses in the DCN is increased after learning, indicating that protein has been added to the postsynaptic density (Weeks *et al.* 2007). This happens within the first 2 or 3 days of learning. The number of excitatory synapses in the DCN that are identifiable by electron microscopy also increases after several days of learning (Kleim *et al.* 2002).

Nothing is yet known about how we determine which cerebellar memories to consolidate. The increases in the size and/or number of synapses in the DCN have interesting implications, though. In forebrain systems, synaptic tagging and capture is a known mechanism for consolidating only the "right" memories. In these systems, physical expansion of the synapse can serve as a tag for synapses that have been modified by short-term memory. CaMKII may be a crucial element for short-term memory, and also may help to tag the synapse; its dephosphorylation without stabilization of the expanded synapse may result in loss of the memory (Redondo & Morris 2011). However, tagging and capture, with all of their associated cellular processes, remain to be established in the cerebellum. It will be interesting to see what discoveries may lie just around the corner.

## INTRINSIC PLASTICITY IN THE DCN

In active neurons, the change in firing that results from a particular synaptic input depends not only on the synapse itself (i.e., transmitter release, transmitter binding, and local ion channels), but on the intrinsic excitability of the postsynaptic cell. Intrinsic excitability is a measure of how easily the cell will fire an action potential. It turns out that intrinsic excitability of DCN neurons is itself plastic (Aizenman & Linden 2000), and changes in excitability could act to amplify changes in synaptic transmission, to consolidate them, or both.

The signals carried by cerebellar neurons, both DCN neurons and the neurons that provide input to them, are ongoing, continuously-changing streams of action potentials. Increased excitability causes the discharge rates in the DCN to increase more in response to a given change in the input, such as an increase in the excitatory input stream or a decrease in the inhibitory input. Intrinsic plasticity can make the DCN neuron more responsive to changes in the signals carried by all of its inputs.

In intrinsic plasticity, burst stimulation of either excitatory or inhibitory inputs to DCN neurons causes a long-lasting decrease in the firing threshold and an increase in input resistance. This means that less depolarizing current is required for the cell to reach its firing threshold. The consequence is increased intrinsic excitability. This intrinsic plasticity requires a rise in intracellular calcium concentration, and takes up to an hour to develop (Zhang *et al.* 2004).

In the vestibular nuclei, hyperpolarizing current injection or the stimulation of P-cell axons causes changes in modulatory BK currents[2] that result in increased intrinsic excitability (Nelson *et al.* 2005). These BK currents are bidirectionally regulated by two kinases, CaMKII and PKC, that phosphorylate the BK channel at different sites (van Welie & du Lac 2011). This means that vestibular neurons have the capacity to increase or decrease their excitability, and that increases and decreases might mask, but would not reverse, each other. A memory that is encoded by BK-channel

[2]BK currents are carried by large-conductance calcium-sensitive potassium channels which normally open after an action potential.

phosphorylation would remain until it is degraded by phosphatase activity. This could confer unusual stability on memory traces in the vestibular nuclei.

Intrinsic plasticity might participate in transferring the memory trace to the DCN during consolidation. Burst stimulation of excitatory and inhibitory afferents to the DCN have a reasonable chance of occurring during motor learning protocols *in vivo*. Climbing fibers evoke complex spikes in P-cells, which are converted to short bursts of action potentials at the P-cell axonal terminals. This means that intrinsic plasticity could happen during motor memory encoding. Hypothetically, consolidation could even be a parallel process that begins at the same time as memory encoding in the cerebellar cortex.

## OTHER CHANGES TO THE MEMORY TRACE

What happens to cerebellar memories that fail to consolidate? The short answer is that we do not know, but we can think about the problem. Behaviorally, we would expect learned motor responses that do not consolidate to be lost. This loss need not correspond to extinction as classically defined, because extinction can occur even after consolidation.

Cerebellar memory *can* be extinguished after consolidation. For example, changes in the vestibulo-ocular reflex can be reversed, apparently an unlimited number of times, simply by removing whatever artificial conditions caused the change. If we then return to those artificial conditions, the learning will occur a second time, with a time course similar to the first time. The fact that we can do this over and over suggests that these memories are truly reversed on the molecular level. The reversal could simply be synaptic plasticity triggered by the new error signal, by some mechanism like parallel-fiber LTP. The learned timing of a blink also shows extinction that involves new synaptic plasticity.

If LTP results in reinsertion of AMPA receptors, replacing those that were lost during LTD, then the "extinguished" memory would be lost, literally without a trace. But sometimes, extinction does not seem to completely erase the memory; some residue remains behind. This phenomenon, known as savings, is detected when the original learning paradigm is reinstated, and it takes less time to learn the association the second time around. Savings in the cerebellum could reflect a prevalence of LTP over LTD at presynaptic parallel-fiber terminals. If LTP causes larger presynaptic changes than LTD does, then LTD cannot fully reverse LTP (Boyden & Raymond 2003). Savings could also reflect a division of the memory trace into cortical and DCN components, which are not equally reversible (Medina *et al.* 2001).

Extinction can occur after memory has consolidated, but an unconsolidated memory can even be lost by a "passive" process like, for example, dephosphorylation of CaMKII by baseline phosphatase activity. Some cerebellar motor memories can be disrupted by stimuli that do not cause learning on their own, but only within a very short time after learning has occurred (Cohen *et al.* 2004). The conditioned response may even decrease while the learning protocol is continued (Ohyama *et al.* 2010). We do not know how these changes work, although there has been some speculation that they involve "decay" of the underlying LTD (Titley *et al.* 2009) and/or a short-term potentiation that itself reverses over time (Ohyama *et al.* 2010). In order to discriminate between different mechanisms occurring at the same synapses, it will be important to study these phenomena in brain slices or cultured cells, where cellular processes can be manipulated. But can cellular processes be linked to phenomena in the real world, such as sensory and motor experience?

## REFERENCES

Ahn, S., Ginty, D.D. & Linden, D.J. (1999) A late phase of cerebellar long-term depression requires activation of CaMKIV and CREB. *Neuron*, **23**, 559–568.

Aizenman, C.D., Manis, P.B. & Linden, D.J. (1998) Polarity of long-term synaptic gain change is related to postsynaptic spike firing at a cerebellar inhibitory synapse. *Neuron*, **21**, 827–835.

Aizenman, C.D. & Linden, D.J. (2000) Rapid, synaptically driven increases in the intrinsic excitability of cerebellar deep nuclear neurons. *Nature Neurosci.*, **3**, 109–111.

Boyden, E.S. & Raymond, J.L. (2003) Active reversal of motor memories reveals rules governing memory encoding. *Neuron*, **39**, 1031–1042.

Cohen, M.R., Meissner, G.W., Schafer, R.J. & Raymond, J.L. (2004) Reversal of motor learning in the vestibulo-ocular reflex in the absence of visual input. *Learn. Mem.*, **11**, 559–565.

Kleim, J.A., Freeman, J.H., Bruneau, R., Nolan, B.C., Cooper, N.R., Zook, A. & Walters, D. (2002) Synapse formation is associated with memory storage in the cerebellum. *Proc. Natl. Acad. Sci. USA*, **99**, 13228–13231.

Linden, D.J. (1996) A protein synthesis-dependent late phase of cerebellar long-term depression. *Neuron*, **17**, 483–490.

McElvain, L.E., Bagnall, M.W., Sakatos, A. & Du Lac, S. (2010) Bidirectional plasticity gated by hyperpolarization controls the gain of postsynaptic firing responses at central vestibular nerve synapses. *Neuron*, **68**, 763–775.

Medina, J.F., Garcia, K.S. & Mauk, M.D. (2001) A mechanism for savings in the cerebellum. *J. Neurosci.*, **21**, 4081–4089.

Muri, R. & Knopfel, T. (1994) Activity induced elevations of intracellular calcium concentration in neurons of the deep cerebellar nuclei. *J. Neurochem.*, **71**, 420–428.

Nelson, A.B., Gittis, A.H. & Du Lac, S. (2005) Decreases in CaMKII activity trigger persistent potentiation of intrinsic excitability in spontaneously firing vestibular nucleus neurons. *Neuron*, **46**, 623–631.

Ohyama, T., Voicu, H., Kalmbach, B.E. & Mauk, M.D. (2010) A decrementing form of plasticity apparent in cerebellar learning. *J. Neurosci.*, **30**, 16993–17003.

Ouardouz, M. & Sastry, B.R. (2000) Mechanisms underlying LTP of inhibitory synaptic transmission in the deep cerebellar nuclei. *J. Neurophysiol.*, **84**, 1414–1421.

Person, A.L. & Raman, I.M. (2010) Deactivation of L-type Ca current by inhibition controls LTP at excitatory synapses in the cerebellar nuclei. *Neuron*, **66**, 550–559.

Pugh, J.R. & Raman, I.M. (2006) Potentiation of mossy fiber EPSCs in the cerebellar nuclei by NMDA receptor activation followed by postinhibitory rebound current. *Neuron*, **51**, 113–123.

Pugh, J.R. & Raman, I.M. (2008) Mechanisms of potentiation of mossy fiber EPSCs in the cerebellar nuclei by coincident synaptic excitation and inhibition. *J. Neurosci.*, **28**, 10549–10560.

Redondo, R.L. & Morris, R.G.M. (2011) Making memories last: The synaptic tagging and capture hypothesis. *Nat. Rev. Neurosci.*, **12**, 17–30.

Smith-Hicks, C., Xiao, B., Deng, R., Ji, Y., Zhao, Z., Shepherd, J.D., Posern, G., Kuhl, D., Huganir, R.L., Ginty, D.D., Worley, P.F. & Linden, D.J. (2010) SRF binding to SRE 6.9 in the Arc promoter is essential for LTD in cultured Purkinje cells. *Nature Neurosci.*, **13**, 1082–1089.

Titley, H.K., Heskin-Sweezie, R. & Broussard, D.M. (2009) Consolidation and disruption of motor memory generalize across stimulus conditions in the vestibulo-ocular reflex. *Brain Res.*, **1267**, 37–43.

van Welie, I. & du Lac, S. (2011) Bidirectional control of BK channel open probability by CAMKII and PKC in medial vestibular nucleus neurons. *J. Neurophysiol.*, **105**, 1651–1659.

Weeks, A.C.W., Connor, S., Hinchcliff, R., LeBoutillier, J.C., Thompson, R.F. & Petit, T.L. (2007) Eye-blink conditioning is associated with changes in synaptic ultrastructure in the rabbit interpositus nuclei. *Learn. Mem.*, **14**, 385–389.

Wulff, P., Schonewille, M., Renzi, M., Viltono, L., Sassoe-Pognetto, M., Badura, A., Gao, Z., Hoebeek, F.E., van Dorp, S., Wisden, W., Farrant, M. & De Zeeuw, C.I. (2009) Synaptic inhibition of Purkinje cells mediates consolidation of vestibulo-cerebellar motor learning. *Nat. Neurosci.*, **12**, 1042–1049.

Zhang, W., Shin, J.H. & Linden, D.J. (2004) Persistent changes in the intrinsic excitability of rat deep cerebellar nuclear neurones induced by EPSP or IPSP bursts. *J. Physiol.*, **561**, 703–719.

Zhang, W. & Linden, D.J. (2006) Long-term depression at the mossy fiber-deep cerebellar nucleus synapse. *J. Neurosci.*, **26**, 6935–6944.

# Section II

## Motor Learning

As we interact with our environment, we continually learn small adjustments to movements. We are not aware of these, but they are important for survival; what would happen if you made an inaccurate movement in reaching for a tree limb to break your fall to the ground? In order to make accurate movements rapidly, we need to go through a process of trial and error (preferably in a less dangerous situation). The cerebellum turns the trial-and-error process into an improvement in the accuracy of the reaching movement.

Motor learning causes a lasting change in a movement, or a new response to a stimulus, that persists after a temporary change in conditions. In humans, establishing fine motor skills—like weighing small quantities in the lab without spilling—is motor learning. A (somewhat different) form of cerebellar-dependent motor learning is present even in larval fish (Aizenberg & Schuman 2011). Learning may well have been a function of the cerebellum from the beginning of its existence.

Motor learning requires some degree of alertness, but it does not require understanding. Even thinking about how walking is done may be enough to disturb our normal gait; it is better not to think about it. As Yogi Berra supposedly said, "Think! How… are you gonna think and hit at the same time?" The same goes for dancing, riding a bicycle, or playing the piano. We listen to the music; visualize where we are going; maybe we think about the sequence of movements, but we certainly do not think about what to do with our left knee joint or index finger, because there is no time for that. Speed is too important.

What exactly is the role of the cerebellum in all this? One possibility is that the cerebellum speeds up skilled movements. It may also make sure that they are timed correctly, with respect to other movements or external events. For example, the cerebellum can learn the precise timing of a protective response (Chapter 5).

Learning always requires the encoding of a memory, as a change somewhere in the brain. To be useful, though, the memory must also be retained. In Chapters 3 and 4, we assumed that both processes take place

*The Cerebellum: Learning Movement, Language, and Social Skills*, First Edition. Dianne M. Broussard.

in the cerebellum, but actually testing this assumption has been difficult. This is partly because the three components of the cerebellar circuit are arranged in a loop: the cortex inhibits the deep nuclei, which inhibit the inferior olive, which has a complex excitatory and feed-forward inhibitory effect on the cortex (Figure 1.4). In other words, a given region of cortex sends a signal that eventually feeds back onto itself. The loop creates a problem for the standard methods that are used to understand the brain's role in behavior. If, for example, we make a lesion in the cerebellar cortex, we disinhibit the DCN, driving DCN neurons into saturation. If we make a lesion in the inferior olive, the cortex does not work normally, and so on. Single-unit recording experiments become difficult to interpret, because whether or not a particular part of the loop (say, the Purkinje cell or P-cell) is the site of learning, it is going to change its response because it is part of the loop. Because of this complexity, it is a good idea to choose a very simple motor task, such as a conditioned reflex, and study it in detail.

# 5 Learning a New Motor Response

Dogs, cats, and many other mammals have a nictitating membrane or "third eyelid" that can protect the cornea; birds and reptiles also have this structure. The rabbit's nictitating membrane response (NMR) is a quick reflex that closes the membrane. The NMR can become associated with different stimuli, and its timing can also be adjusted. For example, if a puff of air to the cornea follows a tone at a consistent interval, a rabbit will learn to time the NMR to occur after the tone begins and just before the air puff occurs. In this protocol the tone is a "conditioned stimulus" and the air puff is the "unconditioned stimulus" (no conditioning is required for a response to an air puff).

It is important to remember that the conditioned NMR is a new response; tones do not normally evoke the NMR. Also, although it is called the "eyeblink reflex" or "aversive blink", we should keep in mind that the NMR is not really a blink. A true blink is closure of the eyelids, which can be voluntary. The NMR, on the other hand, is thought to be completely involuntary. However, true blinks can also be conditioned using the same protocol.

The most commonly used protocols for NMR conditioning are "trace conditioning" and "delay conditioning." For "trace conditioning," which was used in Pavlov's famous experiments, there is an interval (usually, less than a second) between the end of the tone and the beginning of the air puff. Both the cerebellum and the hippocampus are necessary for trace conditioning of NMR responses. In "delay conditioning," the tone is still ongoing when the aversive stimulus begins. The hippocampus is not required for NMR delay conditioning. In fact, human patients with anterograde amnesia can learn an eyeblink response to a tone in a delay protocol. This suggests that in delay conditioning, the memory is acquired using only the cerebellum and without the involvement of forebrain systems. Although changes do happen in the hippocampus during delay conditioning, they are probably irrelevant to the behavioral response. They may have to do with the animal's declarative memory of

*The Cerebellum: Learning Movement, Language, and Social Skills*, First Edition. Dianne M. Broussard.

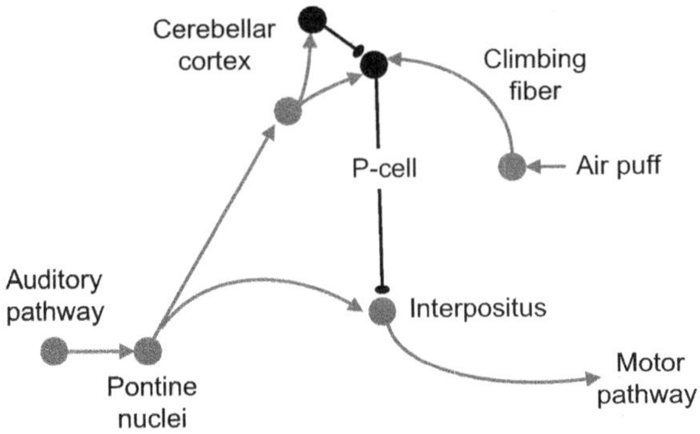


**Figure 5.1** The pathway for NMR learning. Information about the tone arrives on mossy fibers that connect the pontine nuclei with the cerebellar cortex. Collaterals of mossy fibers reach the interpositus, where they converge with inhibitory input from P-cells. Information about the air puff arrives from the trigeminal nerve. Climbing fibers arise from the inferior olive and evoke complex spikes, followed by pauses, in P-cells. Output from the interpositus reaches the red nucleus, part of the motor pathway for the NMR. Gray cells and arrows: excitatory projections. Black cells and round terminals: inhibitory projections.

what occurred (Schmaltz & Theios 1972). From here on, I will focus on delay conditioning.

Delay conditioning of the NMR seems to work best with short daily training sessions repeated over several days. The 24-hour intervals between sessions allow time for memory consolidation. Often, the conditioned response takes several days to become large enough and consistent enough that the animal is considered to be trained. During that time, new memory builds upon previously consolidated memory. This tends to obscure the distinction between learning and consolidation, and it is important to keep that in mind.

For the conditioned NMR, the memory is encoded and stored somewhere in the cerebellum. Signals encoding the conditioned stimulus (the tone) arrive at the cerebellum as excitatory projections from the cochlear nucleus by way of the pontine nuclei (Figure 5.1). These axons have terminals in both the cerebellar cortex and the nucleus interpositus. Signals encoding the air puff or other unconditioned stimulus[1] arrive by way of the inferior olive, on the climbing fibers, which also send collaterals to the interpositus. The same signals also go directly to motor pathways (not shown in the figure), so that the air puff always results in an NMR. This unconditioned response does not require the cerebellum at all.

The site of eyeblink learning could be in the cerebellar cortex or in the nucleus interpositus. But which one? To answer this, the obvious

[1]It is technically easier to use a shock to the orbit ("periorbital shock") instead of an air puff, and this method has become standard. Either way, the stimulus is aversive and painful.

approach is to condition an animal, then make a lesion in the putative storage site to see if the learned response disappears. Almost as soon as these experiments began, signs of trouble appeared. For example, damage to the inferior olive, not considered a likely site for learning, interfered with memory retention (Montarolo *et al.* 1982; Yeo *et al.* 1986). Other early experiments seemed much more definitive, but later were found to have been misinterpreted. Gradually the field is sorting itself out, but the process is far from being complete. In this chapter, I will propose a new interpretation of the NMR results.

## LEARNING IN THE CEREBELLAR CORTEX

The cerebellar cortex is necessary both for learning and for long-term storage of NMR memory. In rabbits, removal of restricted regions of the cortex can abolish the conditioned NMR if the lesion includes a particular region of the posterior lobe (called hemisphere VI, or "HVI"). The effective region included the more medial parts of crus I and crus II, which are in the lateral part of the posterior lobe (Hardiman & Yeo 1992). This could mean that the cortex is the site of memory storage, or that it is necessary for expressing the memory. It is also necessary for learning, because if HVI and the adjacent areas are completely ablated without damaging the interpositus nucleus, the rabbits cannot relearn the NMR (Gruart & Yeo 1995). Mutant mice in which the P-cells have degenerated (pcd mice) have difficulty learning the delay eyeblink task (Chen *et al.* 1996). The eyeblink task is similar to the NMR except that the eyelid is innervated by the facial nerve rather than the abducens. These results could mean that the cortex is the site of learning, or that it supports learning somewhere else.

Loss of the P-cells would probably prevent learning in the interpositus. It would also disinhibit the interpositus, which might impair its capacity to express a memory that is stored there. A better experimental method is to block synapses locally. Blocking synaptic transmission between parallel fibers and P-cells in HVI using CNQX[2] eliminates the NMR to a tone (Attwell *et al.* 1999) and also prevents further learning (Attwell *et al.* 2001). CNQX also blocks most transmission from parallel fibers to the inhibitory interneurons in the cortex. The loss of memory suggests that glutamatergic synapses in the cerebellar cortex participate in storing the memory.

In the CNQX study, timing was not addressed as a separate factor, but it might be a good idea to consider it separately. As already mentioned, there are two requirements for the conditioned NMR: first, the NMR must be generated in response to a different stimulus (the tone); second, the NMR must be timed to reach its peak at the time of the air puff or shock.

[2]CNQX is an antagonist at AMPA receptors. AMPA receptors are thought to be responsible for all excitatory transmission at this synapse.

In the CNQX studies, responses that occurred too early were not considered conditioned responses. This means that the CNQX results support the cortex as the site of learning the correct *timing* of the NMR.

Infusion of lidocaine[3] into the interpositus prevents performance of a previously learned NMR, as well as performance of the conditioned reflex during training, by silencing the interpositus neurons. However, inactivating the interpositus in this way does not prevent learning; the memory is expressed after the lidocaine wears off (Welsh & Harvey 1991). Lidocaine prevents synaptic plasticity in the interpositus by blocking activity in presynaptic terminals and the resulting transmitter release. Since learning occurs anyway, the memory must be encoded upstream, in the cortex. Notably, in this study the rabbits were first conditioned with light as the conditioned stimulus, and this increased the speed of learning after the tone was introduced. All 120 trials of conditioning lasted only a little over 1 hour, so consolidation was probably not a factor.

There is one more key observation to consider: if the P-cells are disconnected from the deep nuclei by injecting picrotoxin[4], learning is completely prevented (Bao *et al.* 2002). What does this mean? One possibility is that inhibitory plasticity at the synapse between the P-cell and the DCN is a mechanism for learning. Alternatively, plasticity in the interpositus, guided by P-cells, might be necessary for the NMR to become large enough to be detected. LTP at the glutamatergic inputs to the interpositus could be induced by the pauses in P-cell activity that are caused by complex spikes.

Based on these results and others, Michael Mauk has proposed that different sites of plasticity are engaged at different phases of learning. Suppose that initially, plasticity in the cerebellar cortex results in a learned timing interval, but the NMR is too small to be measured. Then, once the encoding process has begun, slower plasticity in the nucleus interpositus amplifies the NMR so that it can be detected. Mauk's hypothesis predicts that latent learning of a particular delay by the cerebellar cortex can begin to be expressed after a longer period of training *using a different delay*. This sounds far-fetched, but the prediction was upheld (Ohyama & Mauk 2001). Mauk's hypothesis is especially interesting because it relates to the timing function of the cerebellum.

## CEREBELLAR LTD AND LEARNING

What are the underlying processes in the learning of NMR timing by the cortex? Of the many possibilities, LTD at the PF–PC synapse has attracted

[3]Lidocaine blocks voltage-gated sodium channels, preventing action potentials.

[4]Picrotoxin blocks GABA receptors.

the most attention. Auditory information arrives in the cerebellar cortex as mossy-fiber inputs, which are relayed to P-cells on parallel fibers. Touch signals from the eye, caused by the air puff or periorbital shock, arrive as climbing-fiber inputs. This arrangement means that the air puff could (theoretically) induce LTD at the auditory-input synapses, and naturally it has encouraged many scientists to think of cerebellar LTD as the major mechanism for NMR or eyeblink learning.

If LTD contributes to learning, the patterns of activity in the cerebellum during learning should be consistent with LTD. LTD works best if the climbing fibers are stimulated on every trial just after the P-cell is depolarized (or just after the parallel fibers are stimulated). Complex spikes do occur in response to periorbital shocks, which are delivered after the auditory stimulus (Berthier & Moore 1986; Hesslow & Ivarsson 1994). This means that climbing-fiber activity may, in fact, be paired with depolarization during learning.

Some investigators have interfered with learning by blocking LTD. Mutant mice that are lacking mGluR1 do not have cerebellar LTD, and cannot learn the eyeblink task (Aiba *et al.* 1994). This does not necessarily mean that LTD is the mechanism for eyeblink learning, because other forms of synaptic plasticity could also be affected. In transgenic mice in which protein kinase C is inhibited in P-cells, cerebellar LTD does not occur, and again, other cellular processes besides LTD may be affected. The PKC mutant mice can learn the eyeblink, but the responses are too early, supporting a role for P-cells (but not necessarily LTD) in learned timing.

Even though they do not have LTD, these mice are able to learn the conditioned blink; it is specifically timing that is impaired (Koekkoek *et al.* 2003). But even learned timing may not absolutely require LTD. In a global knockout of the CB1 cannabinoid receptor—which should be lacking LTD—conditioned blinks are small but they do occur, and their timing is normal (Kishimoto & Kano 2006).

As these manipulations become more selective for LTD, they seem to have less effect on eyeblink learning. By specifically targeting the internalization of AMPA receptors in P-cells that is necessary for LTD, a recent study showed that parallel-fiber LTD is not necessary for learning (Schonewille *et al.* 2011). LTD was completely eliminated in the PICK1 knockout and two other knockouts affecting internalization, and eyeblink learning was completely unaffected, at least by any of the measures that have been discussed so far. This single paper may have shifted the balance of opinion against PF–PC LTD as a necessary mechanism for eyeblink learning. While it is still possible that parallel-fiber LTD participates in learning, its role seems to be getting smaller and smaller.

There are, as always, alternatives to this conclusion. If eyeblink learning in the mouse employs a different cerebellar mechanism than NMR learning, or the use of knockouts has generated artifacts[5]—and both of these are possible—then we can still consider LTD is a necessary mechanism for learning timing of the NMR. Or, LTD could be the mechanism for learning timing at an early stage—which is not necessarily detectable—or it could contribute in some other subtle way. The evidence still indicates that the cerebellar cortex is involved in learning of correct timing. In summary, the take-home message seems to be that other forms of cortical plasticity may contribute significantly to memory encoding, and that normally, LTD also contributes. When LTD is absent, other processes take over.

There are two facts that are important to keep in mind at this stage. First, the NMR is a very simple response. There are no joints, much less limbs, involved. Only one movement is required. Only one muscle (per eye) contracts. The vestibulo-ocular reflex, which was also tested, is also extremely simple. Movements that are controlled by the cerebellum can get a lot more complicated, as we will see in later chapters. This means that the NMR probably does not push the cerebellar cortex to the limit of its abilities. Second, the cerebellar cortex is a network with extensive connectivity and, like other networks, it can learn in multiple ways. If LTD is gone it can most likely get around the problem. It is still possible that parallel-fiber LTD would be absolutely required in order for the mouse to learn more complex movement plans.

What other forms of cortical plasticity could contribute to learning? Although CREB phosphorylation is generally considered to be a mechanism for consolidation of memory, it could also participate in memory encoding. The cascade controlling CREB is triggered by noradrenergic receptors. In rats, eyeblink conditioning over several days is impaired by locally blocking either the receptors, or downstream components of the cascade, in the cerebellar cortex and in the interpositus (Cartford *et al.* 2004). With norepinephrine blocked, some conditioned blinks did occur, but they were abnormally small and mistimed, suggesting that the CREB pathway is required for learning the correct timing of the blinks, for amplifying them, and for consolidation.

The cerebellar cortex clearly participates in memory consolidation across sessions of NMR conditioning. Muscimol hyperpolarization of HVI prevents consolidation of the NMR in rabbits, so long as it is done immediately after each training session. The memory from just before the injection is completely lost. Muscimol in the interpositus nucleus does not have the same effect (Attwell *et al.* 2002). All of this probably means that a local consolidation mechanism in the cortex comes into effect shortly after learning.

---

[5]"Knockout" mice cannot express the gene, in this case only in P-cells, and the deficit is life-long. The mice may have compensated in some way during development, causing secondary effects that can only be guessed at.

The details of the local consolidation mechanism are not known, but some hints were obtained using a knockout of calcium/calmodulin-activated protein phosphatase-2B in P-cells. The knockout interferes with both parallel-fiber LTP and intrinsic plasticity while leaving LTD intact. The mice learn correct timing in the eyeblink task, but the memory fails to consolidate (Schonewille *et al.* 2010). These results favor LTP and intrinsic plasticity as mechanisms for cortical memory consolidation.

Some local consolidation processes may be triggered by LTD. For example, the P-cells in HVI have increased intrinsic and synaptic excitability that is retained 24 hours after NMR conditioning. Furthermore, LTD is occluded (indicating that it has happened). Perhaps the increased excitability can homeostatically compensate for LTD, or vice versa (Schreurs *et al.* 1997).

So far, it appears that mechanisms in the cerebellar cortex, including at least some forms of synaptic plasticity, are important for eyeblink learning and for memory consolidation. The exact nature of these processes is still being established. But what about long-term storage?

## THE ENGRAM FOR THE NMR

The long-term memory trace must somehow contain the information needed to generate the timed NMR. Since the NMR is a new response, we need records both for generating the response and for its timing. Are the two records stored in the same place? Does the memory stay where it is first formed, or does it move? If there are two sites, exactly what information is contained in each one?

Damage to the inferior olive interferes with memory retention (Montarolo *et al.* 1982; Yeo *et al.* 1986), probably because it triggers extinction. Here is how it works: whenever the rabbit blinks, including during a learned NMR, the interpositus generates a burst of action potentials. Climbing-fiber activity is reduced during the burst, because the inferior olive is inhibited by the interpositus. Blocking inhibitory synapses in the inferior olive interrupts this pathway and prevents extinction of the NMR (Medina *et al.* 2002). Extinction also requires either sensory feedback or reafference, signaling that the NMR has occurred (Krupa & Thompson 2003). The fact that climbing fibers are required to be silent suggests a cerebellar cortical mechanism for extinction. In fact, extinction requires the participation of both cortex and interpositus, because injecting muscimol[6] into the interpositus also prevents it (Rammani & Yeo 1996).

---

[6]Muscimol is a selective agonist for the $GABA_A$ receptor, which is present at most inhibitory synapses in the brain. It reduces the activity of the neurons in any area where it is injected, and often silences them.

The requirements for extinction can inform us about long-term memory storage. For example, after NMR learning in rabbits, lesions of the cerebellar cortex that destroy timing information also prevent extinction of the (mistimed) NMR. The rabbit is left with a useless, early NMR that will not go away (Perrett & Mauk 1995). In a sense, cortical lesions throw away the key to the memory, but they do not destroy it. This would be consistent with long-term memory storage in the interpositus nucleus.

Storage in the interpositus was supported by an early study, in which the infusion of muscimol in either the cerebellar cortex or the interpositus was reported to prevent eyeblink learning (Krupa *et al.* 1993). This seemed to mean that the interpositus was the site of learning. But the trials were spaced over several days, which allowed time for memory consolidation between the trials. This means that consolidation could account for the effect of muscimol. Furthermore, we already know that inactivating the interpositus does not prevent memory encoding (Welsh & Harvey 1991). Muscimol can have very long-lasting effects and might prevent memory transfer, resulting in the loss of new memory that was stored transiently in the cortex.

Lesions of the interpositus do abolish the long-term memory for the NMR response to a tone. The key region of the interpositus receives input from the anterior lobe as well as HVI. But the effects of interpositus lesions cannot prove that long-term memory is stored in the interpositus. They cannot even rule out memory storage in the cortex exclusively, because signals from the cortex must pass through the interpositus on their way out of the cerebellum. It is more interesting that large lesions of the cerebellar cortex can be made *without* abolishing the consolidated NMR (although its timing may be affected) (McCormick & Thompson 1984; Lavond *et al.* 1985). Furthermore, memory consolidation is unaffected by hyperpolarizing HVI 8 hours after injection (Cooke *et al.* 2004). One possibility is that the memory that generates the NMR to a tone has been transferred out of the cortex by that time.

In summary, it's clear that the interpositus, like the cortex, participates in memory consolidation. In the transfer hypothesis, memory is consolidated by being transferred from the cortex to the interpositus and re-encoded there. In Mauk's hypothesis, consolidation might operate on a memory that was originally formed in the interpositus, and no transfer would be necessary. Either way, the memory would end up in the interpositus.

The interpositus has the capacity to learn, as outlined in Chapter 4. Under extreme conditions, the interpositus can even learn the NMR without P-cells. After their P-cells have degenerated, pcd mice can very slowly learn a less-vigorous blink than normals (Chen *et al.* 1996). As Figure 5.1 illustrates, signals for the NMR pass through the anterior interpositus nucleus and can bypass the cortex (Ohyama *et al.* 2006). Even in normal animals, during learning, the amplitude of the movement of the nictitating membrane increases, and this change could be mediated by plasticity in the glutamatergic pathway through the interpositus. LTP of glutamatergic inputs to the interpositus, triggered by the release of P-cell inhibition, could be the mechanism for memory transfer or for the changes in amplitude, or both. If transfer occurs, it probably occurs within 24 hours, because by that time, some memory is present in the interpositus (Medina *et al.* 2001).

In the transfer scenario, synaptic plasticity in the interpositus could be triggered by the events that occur during a recently learned NMR. P-cells may stop firing at the correct time for the NMR, even if it is not large enough to be detected. During the P-cell pause, the mossy fibers that encode the tone (or other CS) should still be active, providing the exact conditions that are required for LTP of the mossy-fiber inputs to interpositus neurons. LTP of the inputs signaling the tone could then amplify the NMR. LTP of either excitatory or inhibitory inputs to interpositus neurons can then trigger increases in excitability which may further amplify the conditioned response (Zheng & Raman 2010).

Some information remains stored in the cortex, even after consolidation; for example, the timing information may remain stored there. After damage to the cortex of the anterior lobe, the NMR may lose its accurate timing; it tends to be too early as well as smaller (Perrett *et al.* 1993; Garcia *et al.* 1999). Also, if the P-cells are disconnected from the deep nuclei by injecting GABA antagonists after learning, the learned responses occur too early and are over before the shock (Garcia & Mauk 1998; Bao *et al.* 2002). All of this is consistent with the storage of timing information in the cortex.

After extinction has occurred, the NMR can be relearned quickly, suggesting that extinction does not reverse learning but somehow masks the memory. Also, if picrotoxin is injected into the interpositus following extinction, isolating it from the cortex, the NMR reappears. The re-emergent NMR is not correctly timed but occurs too early, right after the tone onset (Medina *et al.* 2001). This is presumably because response timing remains stored in the cortex, while the long-term memory of NMR performance to a tone is stored in the interpositus. But there is something else to be learned from this result: extinction may be a cortical process, so that injecting picrotoxin obscures its effects, revealing the consolidated memory that still is stored in the interpositus. Interestingly, the duration of extinction training does not affect this result; it can go on for weeks, and still not reverse the interpositus memory trace (Medina *et al.* 2001). By analogy to extinction, then, this implies that memory encoding is also be a cortical process exclusively.

Processes involving protein synthesis in the interpositus are necessary for retention of the NMR over time (Bracha *et al.* 1998). After several days of eyeblink conditioning, there is an increased number of excitatory synapses per neuron in the interpositus (Kleim *et al.* 2002), suggesting that new synapses may be formed in the interpositus that allow novel stimuli such as tones to generate timed eyeblinks. It appears that potentiation may happen first, and new synapses then form a bit later on. For example, after 2–3 days of conditioning, the number of synapses had not yet increased significantly, but the length of excitatory synapses had increased, suggesting that they had potentiated (Weeks *et al.* 2007). Increased length of the postsynaptic density is thought to be a correlate of the "maintenance phase" of LTP.

## CONCLUSIONS

In Chapter 4, I suggested that the cerebellar cortex may be the first site of memory encoding, and that consolidation may rely on re-encoding of the memory in the DCN. After examining the evidence from the NMR system, this scenario is still plausible, but it is not the whole story. Mauk's hypothesis, which requires two sites for memory encoding, one for generating the NMR and another for timing it, is more complex but it may provide a better fit to the data. There is significant support for this hypothesis.

But there is still significant doubt. In fact, there is major disagreement in this field about the site for memory encoding, the site for memory storage, and the mechanisms for learning and consolidation. Some of the confusion has arisen because the NMR is a new motor task, and takes some time to appear. Because in many experiments only large responses are detectable, it can take days to see a response that meets the criterion. This means that the early stages of memory encoding and rapid consolidation are below the radar, so to speak. Most of the learning that can be studied has already undergone consolidation. As learning proceeds, transfer is taking place and probably new encoding as well. Fortunately, there are learning tasks that do not have these problems, and some of them will be described in the next chapter.

## REFERENCES

Aiba, A., Kano, M., Chen, C., Stanton, M.E., Fox, G.D., Herrup, K., Zwingman, T.A. & Tonegawa, S. (1994) Deficient cerebellar long-term depression and impaired motor learning in mGluR1 mutant mice. *Cell*, **79**, 377–388.

Aizenberg, M. & Schuman, E.M. (2011) Cerebellar-dependent learning in larval zebrafish. *J. Neurosci.*, **31**, 8708–8712.

Attwell, P.J.E., Rahman, S., Ivarsson, M. & Yeo, C.H. (1999) Cerebellar cortical AMPA-kainate receptor blockade prevents performance of classically conditioned nictitating membrane responses. *J. Neurosci.*, **19**, RC45.

Attwell, P.J.E., Rahman, S. & Yeo, C.H. (2001) Acquisition of eyeblink conditioning is critically dependent on normal function in cerebellar cortical lobule HVI. *J. Neurosci.*, **21**, 5715–5722.

Attwell, P.J.E., Cooke, S.F. & Yeo, C.H. (2002) Cerebellar function in consolidation of a motor memory. *Neuron*, **34**, 1011–1020.

Bao, S., Chen, L., Kim, J.J. & Thompson, R.F. (2002) Cerebellar cortical inhibition and classical eyeblink conditioning. *Proc. Natl. Acad. Sci. USA*, **99**, 1592–1597.

Berthier, N.E. & Moore, J.W. (1986) Cerebellar Purkinje cell activity related to the classically conditioned nictitating membrane response. *Exp. Brain Res.*, **63**, 341–350.

Bracha, V., Irwin, K.B., Webster, M.L., Wunderlich, D.A., Stachowiak, M.K. & Bloedel, J.R. (1998) Microinjections of anisomycin into the intermediate cerebellum during learning affect the acquisition of classically conditioned responses in the rabbit. *Brain Res.*, **788**, 169–178.

Cartford, M.C., Samec, A., Fister, M. & Bickford, P.C. (2004) Cerebellar norepinephrine modulates learning of delay classical eyeblink conditioning: Evidence for post-synaptic signaling via PKA. *Learn. Mem.*, **11**, 732–737.

Chen, L., Bao, S., Lockard, J.M., Kim, J.J. & Thompson, R.F. (1996) Impaired classical eyeblink conditioning in cerebellar-lesioned and Purkinje cell degeneration (pcd) mutant mice. *J. Neurosci.*, **16**, 2829–2838.

Cooke, S.F., Attwell, P.J.E. & Yeo, C.H. (2004) Temporal properties of cerebellar-dependent memory consolidation. *J. Neurosci.*, **24**, 2934–2041.

Garcia, K.S. & Mauk, M.D. (1998) Pharmacological analysis of cerebellar contributions to the timing and expression of conditioned eyelid responses. *Neuropharmacology*, **37**, 471–480.

Garcia, K.S., Steele, P.M. & Mauk, M.D. (1999) Cerebellar cortex lesions prevent acquisition of conditioned eyelid responses. *J. Neurosci.*, **19**, 10940–10947.

Gruart, A. & Yeo, C.H. (1995) Cerebellar cortex and eyeblink conditioning–bilateral regulation of conditioned responses. *Exp. Brain Res.*, **104**, 431–438.

Hardiman, M.J. & Yeo, C.H. (1992) The effect of kainic acid lesions of the cerebellar cortex on the conditioned nictitating membrane response in the rabbit. *Eur. J. Neurosci.*, **4**, 966–980.

Hesslow, G. & Ivarsson, M. (1994) Suppression of cerebellar Purkinje cells during conditioned responses in ferrets. *Neuroreport*, **5**, 649–652.

Kishimoto, Y. & Kano, M. (2006) Endogenous cannabinoid signaling through the CB1 receptor is essential for cerebellum-dependent discrete motor learning. *J. Neurosci.*, **26**, 8829–8837.

Kleim, J.A., Freeman, J.H., Bruneau, R., Nolan, B.C., Cooper, N.R., Zook, A. & Walters, D. (2002) Synapse formation is associated with memory storage in the cerebellum. *Proc. Natl. Acad. Sci. USA*, **99**, 13228–13231.

Koekkoek, S.K.E., Hulscher, H.C., Dortland, B.R., Hensbroek, R.A., Elgersma, Y., Ruigrok, T.J.H. & De Zeeuw, C.I. (2003) Cerebellar LTD and learning-dependent timing of conditioned eyelid responses. *Science*, **301**, 1736–1739.

Krupa, D.J., Thompson, J.K. & Thompson, R.F. (1993) Localization of a memory trace in the mammalian brain. *Science*, **260**, 989–992.

Krupa, D.J. & Thompson, R.F. (2003) Inhibiting the expression of a classically conditioned behavior prevents its extinction. *J. Neurosci.*, **23**, 10577–10584.

Lavond, D.G., Hembree, T.L. & Thompson, R.F. (1985) Effect of kainic acid lesions of the cerebellar interpositus nucleus on eyelid conditioning in the rabbit. *Brain Res.*, **326**, 179–182.

McCormick, D.A. & Thompson, R.F. (1984) Cerebellum: essential involvement in the classically conditioned eyelid response. *Science*, **223**, 296–299.

Medina, J.F., Garcia, K.S. & Mauk, M.D. (2001) A mechanism for savings in the cerebellum. *J. Neurosci.*, **21**, 4081–4089.

Medina, J.F., Nores, W.L. & Mauk, M.D. (2002) Inhibition of climbing fibres is a signal for the extinction of conditioned eyelid responses. *Nature*, **416**, 330–333.

Montarolo, P.G., Palestini, M. & Strata, P. (1982) The inhibitory effect of the olivocerebellar input on the cerebellar Purkinje cells in the rat. *J. Physiol.*, **332**, 187–202.

Ohyama, T. & Mauk, M.D. (2001) Latent acquisition of timed responses in cerebellar cortex. *J. Neurosci.*, **21**, 682–690.

Ohyama, T., Nores, W.L., Medina, J.F., Riusech, F.A. & Mauk, M.D. (2006) Learning-induced plasticity in deep cerebellar nucleus. *J. Neurosci.*, **26**, 12656–12663.

Perrett, S.P., Ruiz, B.P. & Mauk, M.D. (1993) Cerebellar cortex lesions disrupt the learning-dependent timing of conditioned eyelid responses. *J. Neurosci*, **13**, 1708–1718.

Perrett, S.P. & Mauk, M.D. (1995) Extinction of conditioned eyelid responses requires anterior lobe of cerebellar cortex. *J. Neurosci.*, **15**, 2074–2080.

Rammani, N. & Yeo, C.H. (1996) Reversible inactivations of the cerebellum prevent the extinction of conditioned nictitating membrane responses in rabbits. *J. Physiol.*, **495**, 159–168.

Schmaltz, L.W. & Theios, J. (1972) Acquisition and extinction of a classically conditioned response in hippocampectomized rabbits *(Oryctolagus cuniculus)*. *J. Comp. Physiol. Psychol.*, **79**, 328–333.

Schonewille, M., Belmeguenai, A., Koekkoek, S.K.E., Houtman, S.H., Boele, H.J., van Beugen, B.J., Gao, Z., Badura, A., Ohtsuki, G., Amerika, W.E., Hosy, E., Hoebeek, F.E., Elgersma, Y., Hansel, C. & De Zeeuw, C.I. (2010) Purkinje cell-specific knockout of the protein phosphatase PP2B impairs potentiation and cerebellar motor learning. *Neuron*, **67**, 618–628.

Schonewille, M., Gao, Z., Boele, H.J., Veloz, M.F.V., Amerika, W.E., Simek, A.A.M., De Jeu, M.T.G., Steinberg, J.P., Takamiya, K., Hoebeek, F.E., Linden, D.J., Huganir, R.L. & De Zeeuw, C.I. (2011) Reevaluating the role of LTD in cerebellar motor learning. *Neuron*, **70**, 43–50.

Schreurs, B.G., Tomsic, D., Gusev, P.A. & Alkon, D.L. (1997) Dendritic excitability microzones and occluded long-term depression after classical conditioning of the rabbit's nictitating membrane response. *J. Neurophysiol.*, **77**, 86–92.

Weeks, A.C.W., Connor, S., Hinchcliff, R., LeBoutillier, J.C., Thompson, R.F. & Petit, T.L. (2007) Eye-blink conditioning is associated with changes in synaptic ultrastructure in the rabbit interpositus nuclei. *Learn. Mem.*, **14**, 385–389.

Welsh, J.P. & Harvey, J.A. (1991) Pavlovian conditioning in the rabbit during inactivation of the interpositus nucleus. *J. Physiol.*, **444**, 459–480.

Yeo, C.H., Hardiman, M.J. & Glickstein, M. (1986) Classical conditioning of the nictitating membrane response of the rabbit. IV. Lesions of the inferior olive. *Exp. Brain Res.*, **63**, 81–92.

Zheng, N. & Raman, I.M. (2010) Synaptic inhibition, excitation, and plasticity in neurons of the cerebellar nuclei. *Cerebellum*, **9**, 56–66.

## FURTHER READING

Mauk, M.D., Garcia, K.S., Medina, J.F. & Steele, P.M. (1998) Does cerebellar LTD mediate motor learning? Toward a resolution without a smoking gun. *Neuron*, **20**, 359–362.

Medina, J.F., Repa, J.C., Mauk, M.D. & LeDoux, J.E. (2002) Parallels between cerebellum- and amygdala-dependent conditioning. *Nature Rev. Neurosci.*, **3**, 122–131.

De Zeeuw, C. & Yeo, C.H. (2005) Time and tide in cerebellar memory formation. *Curr. Opinion in Neurobiol.*, **15**, 667–674.

# 6 Recalibration for Fine Motor Control

Fine motor control establishes very precise timing; it also adjusts the amplitude and speed of movement. It is difficult to measure the amplitude and speed of the eyeblink reliably, but for actual eye movements, it is simpler. Because eye movements are almost pure rotations, we can measure them precisely and accurately using search coils, which detect the exact orientation of the eye within a magnetic field by measuring the induced current. In Chapter 6, I will make use of eye-movement systems to illustrate how cerebellar motor learning can provide fine adjustments, increasing or decreasing the speed and amplitude of movements to make them more accurate.

The best-studied eye-movement system, with regard to learning, is the vestibulo-ocular reflex (VOR). The VOR is such a simple system that it is not terribly interesting in itself (except, of course, to its dedicated fans). But there is a payoff for understanding the VOR: building on what we learn from this simple system, we can move on to understand other, more complex forms of motor control and motor learning.

## A STABLE PLATFORM FOR VISION

As anyone knows who has tried to use binoculars from a moving vehicle, the visual system is not very good at stabilizing gaze. Fortunately, the VOR does not rely on vision, but on the inner ear. Its calibrated "feed-forward" pathway stabilizes gaze using signals from the vestibular nerve (Figure 6.1).[1] The VOR moves the eyes at exactly the right speed to counteract head movement, so that we can see during movement, including

[1]The vestibular nerve is a subdivision of the eighth cranial nerve, which is usually called the auditory nerve. Sensory fibers that innervate the three semicircular canals, the saccule, and the utricle encode different aspects of head movement, and travel in different branches of the vestibular nerve. The cell bodies of these sensory fibers are located in the vestibular ganglion.

*The Cerebellum: Learning Movement, Language, and Social Skills*, First Edition. Dianne M. Broussard.

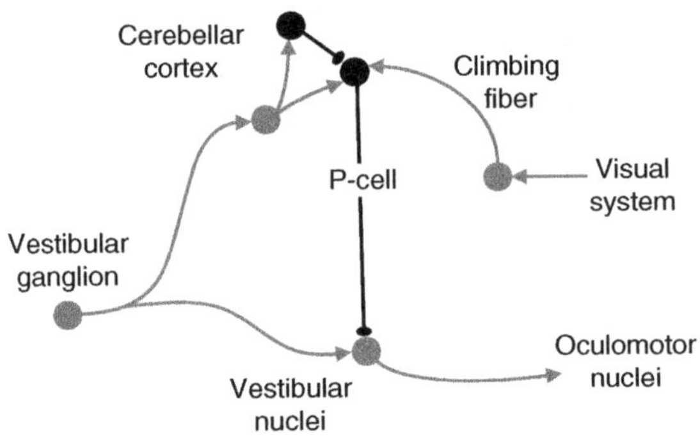


**Figure 6.1** A simplified version of the VOR circuit and its connection to the cerebellum. The VOR stabilizes gaze by a direct pathway from the vestibular ganglion through the vestibular nuclei to motoneurons that move the eyes. The flocculus provides a side loop, including P-cells that also project to the vestibular nuclei. The visual input comes from a short brainstem pathway. Gray cells and arrows, excitatory projections; black cells and round terminals, inhibitory projections.

random bumps, bounces, and vibrations. We are not even aware of the VOR as long as it works correctly. We mostly notice when it is inaccurate, because we feel unsteady or seasick.

The neurons that generate the VOR are located in the vestibular nuclei of the medulla. The vestibular nerve, vestibular nuclei, and oculomotor nuclei are connected in several short reflex pathways that can respond quickly to all head movements. The VOR works well both at low frequencies of rotation[2] (less than 1 Hz) and at much higher frequencies (up to 30 Hz). Calibration of the VOR depends on the cerebellar cortex. Purkinje cells (P-cells) in the flocculus and ventral paraflocculus project directly to the vestibular nuclei.

Vestibular neurons are like DCN neurons in that they have specialized ion channels that allow them to fire spontaneously. There are also some differences; for example, some vestibular neurons that project to motoneurons do not receive any input from the cerebellar cortex, whereas in the DCN, all projection neurons receive cortical input. Lobules IX and X of the cerebellum—the vestibulocerebellum—are also different from the rest of the cerebellar cortex. They have some unique neurons (the unipolar brush cells), and their P-cells have unique, specialized ion channels (Kim *et al.* 2012 ). Both of these features expand the linear range of the "vestibular" P-cells, so that they can generate a linear response to a variety of input signals.

[2]"Frequency of rotation" refers to the frequency of a sinusoidal, back-and-forth movement. Sinusoids are standard stimuli for testing of the VOR, both because they are simple to generate and because they are analytically useful.

The VOR is calibrated by motor learning, which corrects inaccuracies in the gain of the reflex pathway. Gain is a property of linear systems; it is the constant of proportionality that relates the output to the input. For the VOR, "gain" refers to the ratio of the angular velocity of the output signal (eye movement) to that of the input signal (head movement). Motor learning adjusts the speed of the eye movement so that it perfectly compensates for sudden movements of the head. Depending on conditions, learning can either increase or decrease the gain of the VOR. The VOR must learn in any situation where head motion reliably causes the visual scene to move on the retina—for example, if muscles become weak or there is vestibular damage. Such motion of the image on the retina is called "retinal slip."

Climbing fibers carry a visual signal to P-cells in the vestibulocerebellum (Maekawa & Simpson 1973). This fact has led to a widespread belief that cerebellar LTD is the mechanism for VOR motor learning, as was originally proposed by Masao Ito (1972). But nonvisual signals can also trigger learning. Any time that our gaze consistently moves in space while the head moves, learning occurs. Furthermore, on the cellular level, there are many other candidates for learning mechanisms besides LTD.

## *Memory encoding by P-cells*

The circuit in Figure 6.1, which includes the flocculus, is required for VOR motor learning. Permanent lesions of the flocculus in rabbits (Nagao 1983), or of both the flocculus and the ventral paraflocculus in primates (Rambold *et al.* 2002), completely prevent learning. This does not necessarily mean that the memory is encoded in the cerebellar cortex, because the cerebellar cortex could also be required for learning that takes place downstream, in the vestibular nuclei.

In the laboratory, learning can be induced in the VOR by magnifying or otherwise manipulating vision. Magnified vision causes images to move on the retina whenever the head moves. This retinal slip is conveyed to the flocculus by the climbing fibers, and is a useful error signal. The climbing-fiber signal is distinct for situations where the VOR fails to stabilize gaze on the visual scene. The mossy fibers and parallel fibers in the flocculus respond directly to the inner ear signal (see Figure 6.1), and the climbing fibers respond to the retinal slip, but after a delay of 100 ms or so. This sequence of events is similar to the relationship between unconditioned and conditioned stimuli for eyeblink learning. Mossy fibers also convey other error signals, such as tracking eye motion, which can cause the VOR to learn a new gain.

For low frequencies of rotation (less than 2 Hz), the same signals also converge in the vestibular nuclei and could guide learning there; but this is not the case for high frequencies, where the cerebellar cortex seems to be the only possible site for learning (Raymond & Lisberger 1998). The high frequencies—2 Hz and higher—are especially important, because they make up the sudden, unpredictable movements that the VOR must compensate for. High frequencies are present in natural head movements. The flocculus is important for all VOR learning, but its role may vary. At low frequencies, temporary inactivation of the flocculus using lidocaine just after learning seems to inactivate the memory (McElligott *et al.* 1998; Anzai *et al.* 2010). Blockade of excitatory synapses in the flocculus has the same effect (Kassardjian *et al.* 2005).

It is clear that the flocculus dominates memory encoding, if—and this is an important "if"—all of the cerebellar cortical mechanisms for learning are in play. For example, cerebellar LTD may encode increases in VOR gain. Many P-cells respond to rotation by increasing their discharge rates along with the vestibular neurons that they inhibit (Figure 6.1). Decreasing the responses of floccular P-cells to rotation might actually *increase* the responses of those vestibular neurons.

A decrease in P-cell responsiveness could be accomplished by LTD of inputs from particular mossy fibers, that is, those that respond to head rotation with a polarity that matches the P-cell's response. In support of this idea, blockade of mGluR1 prevents gain increases (but not decreases) in a protocol using magnified vision (Titley *et al.* 2010).

On the other hand, it is clear that LTD is not the only mechanism for learning of gain increases. Selectively disabling parallel-fiber LTD, by targeting internalization of AMPA receptors in P-cells, does not affect VOR motor learning at low frequencies (Schonewille *et al.* 2011). This means that other learning mechanisms (such as plasticity in the vestibular nuclei) may be brought into play at low frequencies, if LTD is not available.

The work on the VOR illustrates one important point very clearly: different mechanisms can come into play for different adjustments of gain. For example, we have only a limited ability to increase the VOR's gain by learning, but gain decreases are easier to learn, and are also more stable. Under some circumstances, learned decreases in gain are difficult to extinguish (Boyden & Raymond 2003).

There may be other mechanisms available, and possibly more sites involved, for gain decreases than for increases. Gain decreases do not require nitric oxide, mGluR1 receptors, cerebellar LTD (Schonewille *et al.* 2011), or postsynaptic cerebellar LTP (Schonewille *et al.* 2010). Either mGluR1 blockade, or selective disabling of cerebellar LTP and intrinsic plasticity, will actually cause the gain to decrease in the protocol that is meant to cause increases (Schonewille *et al.* 2010; Titley *et al.* 2010). Gain increases (at low frequencies) are still possible if only LTD is impaired, but LTP seems to be required instead (Schonewille *et al.* 2011).

How can both cerebellar LTD and LTP participate in increasing the gain of the VOR? To understand this, we need to realize that different P-cells in the flocculus respond to eye movements that are in different directions. Depending on the polarity of the granule-cell and P-cell responses with respect to head movement, they could cause an increase in VOR gain by either increasing or decreasing their responsiveness. Either way, the reverse pattern should cause a decrease in gain. What this means is that the participation of both LTP and LTD in both directions of VOR motor learning might actually make the best use of the population of P-cells, by including cells whose responses have different polarities. One possibility is that during learning, LTP first comes into play to increase the active synapses and the number of signals available, and then LTD prunes these into an optimal configuration (Gao *et al.* 2012).

If the flocculus is the primary site for learning, it may also be a site for long-term memory. Should we look for appropriate changes in signaling in the flocculus, after gain increases have been learned? One would think so, but there is a problem. Like other regions of the cerebellum, the flocculus is part of a loop—in fact, at least two loops. Not only is there feedback from the vestibular nuclei via the inferior olive, but in the VOR pathway there is also excitatory feedback from the vestibular nuclei to the cerebellar cortex of the flocculus and ventral paraflocculus. What is fed back is an efference copy of the motor signals. The excitatory feedback, reaches the very same P-cells that are the putative sites of motor learning. Therefore when a new gain has been learned, and the eye-movement signals change their amplitude, this automatically changes the P-cells' discharge patterns.

The responses of P-cells during very abrupt head movements do change when memory is encoded, but the P-cells are not entirely responsible for long-term storage of the change in the VOR. The VOR is a very quick reflex; under ideal conditions it can respond in less than one hundredth of a second (Broussard *et al.* 1992). At least some of the long-term memory for the gain of this quick response is stored in the vestibular nuclei, in a direct pathway (Lisberger *et al.* 1994b). Even though the cerebellum is a timing device, the P-cell responses are too sluggish to be useful in this context (Lisberger *et al.* 1994a).

## Other sites of plasticity

Although the parallel-fiber inputs to P-cells are an important site of memory encoding, they are not the only site. If climbing-fiber signaling is eliminated by carefully controlling the experimental conditions, the VOR can still learn a gain decrease at a low frequency (Ke *et al.* 2009). At low frequencies, both simple spikes and complex spikes in the P-cells are appropriate to guide learning independently: the simple spikes could guide learning in the vestibular nuclei, while the complex spikes guide learning via LTD in the cerebellar cortex. If they do, this predicts that in situations where both signals are present, their contributions to learning should be additive. This prediction is upheld (Ke *et al.* 2009), indicating that both P-cells and vestibular neurons are sites of learning.

Even within the cerebellar cortex, P-cells are not the only sites where memory is encoded. Other sites, such as the glomeruli of the granular layer, also could contribute to learning in the absence of complex-spike error signals. In situations where the climbing fibers do signal errors, they are not restricted to causing learning at P-cell inputs, but may also trigger learning in the vestibular nuclei and at the synaptic inputs to the molecular-layer interneurons. Plasticity at the glomeruli and plasticity in the molecular layer may interact synergistically to adjust the response amplitudes for P-cells with various on-directions (Gao *et al.* 2012).

The possibility that the vestibular nuclei provide a site or sites for memory encoding has been impossible to investigate using destructive lesions, temporary inactivation, or GABA antagonists, because these manipulations stop the VOR from working. However, the vestibular nuclei do meet other criteria for a site of learning. For a start, in the vestibular nuclei, the vestibular signals converge with a gaze-velocity signal that is carried by P-cells. Gaze velocity is the movement of the eyes with respect to the outside world. If the VOR works perfectly, then gaze velocity is zero; if gaze velocity is not zero, then it is time for the VOR to learn. There is one exception: the gaze-velocity signal carried by P-cells fails at high frequencies (which make up most natural head movements), so it seems likely that VOR gain at high frequencies is adjusted only in the cerebellar cortex (Raymond & Lisberger 1998).

There are cellular mechanisms in the vestibular nuclei that can support learning; in fact, the synapses in the short reflex pathway of the VOR itself exhibit both LTP and LTD. In situations where the gain of the VOR needs to increase, P-cell input to the interneurons in the reflex coincides with their excitatory synaptic input from the vestibular ganglion. This situation induces LTP in the excitatory synapses, by a mechanism that involves calcium-permeable AMPA receptors. But in situations where the gain needs to decrease, P-cell input does not exactly coincide with excitatory synaptic input, and LTD is induced by a default mechanism requiring calcium entry through NMDA receptors (McElvain *et al.* 2010). In other words, LTD does not require the P-cell input in order to occur.

In summary, because of the wide variety of mechanisms at its disposal, the VOR is able to learn without cerebellar LTD and LTP, and even without climbing-fiber signals.

## The engram for the VOR

Like memory in the NMR, memory in the VOR consolidates over time. We can easily disrupt newly formed memory, if we do it within 90 minutes or so after the start of learning (Titley *et al.* 2007). Late-phase cerebellar LTD could contribute to consolidating memory for the VOR. Calcium/calmodulin-dependent protein kinase IV (CaMKIV), which is important for late-phase LTD, is also necessary for long-term memory retention. In particular, CaMKIV is necessary for retention of gain increases at high frequencies, which may be the type of VOR motor learning that is most reliant on parallel-fiber LTD for encoding (Boyden *et al.* 2006). Stellate and basket cells, which release GABA, may also play an important part in consolidation, because mice that do not have functional GABA receptors on their P-cells are unable to consolidate VOR motor memory. Interestingly, in this last case the deficit is more general, affirming the important role of inhibitory interneurons in consolidating both gain increases and gain decreases (Wulff *et al.* 2009).

In the long term, memory for the VOR appears to be transferred out of the cerebellar cortex, at least to some extent. As already described, reversibly inactivating the cortex, or blocking cortical synapses, prevents learning and reverses recent changes in VOR gain (that is, changes learned less than 1 hour previously). However, these manipulations cannot reverse changes in VOR gain that were learned over several days (Luebke & Robinson 1994; Kassardjian *et al.* 2005; Anzai *et al.* 2010). In goldfish, it is possible to remove the cerebellar cortex during the experiment and if this is done after a few hours of learning, part of the memory remains. In fact, the short-latency response to an abrupt head movement (which contains high frequencies) retains its learned amplitude perfectly. This suggests that the memory began to be transferred out of the cortex within a few hours after it was first encoded (Pastor *et al.* 1994). LTP and LTD of the excitatory synapses in the vestibular nuclei provide two possible consolidation mechanisms. In addition, intrinsic excitability of vestibular neurons can be altered by their P-cell inputs in a form of intrinsic plasticity that could amplify or consolidate VOR gain changes (Nelson *et al.* 2005).

Using the VOR, we have established some rules that may also hold for other types of motor learning, especially for adjustments to the amplitude and speed of movements. The rules are as follows: First, some component of memory is encoded in the cerebellar cortex under the guidance of climbing fibers. Second, LTD and LTP at the parallel-fiber input to P-cells contribute to memory encoding when they are available, but other cortical mechanisms can also contribute. Third, some component of memory is transferred to the vestibular nuclei (or DCN). Fourth, plasticity in the vestibular nuclei (or DCN) can contribute to memory encoding in particular situations.

## *Is the VOR like the NMR?*

The NMR and the VOR share some common features of learning, but some differences arise from the tasks themselves. For one thing, in the eyeblink/NMR system, what is being learned is the timing of the response. Learned timing is encoded and, apparently, stored for the long term in the cerebellar cortex, while amplitude is stored for the long term in the interpositus. On the other hand, the optimal delay for the VOR is always zero. Timing in the VOR is often adjusted during motor learning, but this appears to be a side effect of changing the gain of particular components of the pathway. For example, decreasing the gain of the VOR might result in a slightly brisker response, and increasing the gain might slow down the response, depending on the frequency. The fact that these effects are frequency-dependent suggests that they are side effects of altering transmission through adaptive linear filters (Kimpo *et al.* 2005).[3] Linear filters cannot generate a delay that does *not* depend on frequency. But that is precisely what the NMR learns: a delay that does not depend on stimulus frequency. The presence of linear filters in the VOR might be a fundamental difference between the cerebellar modules that control the NMR and the VOR.

For the VOR as for the NMR, amplitude (i.e., gain) is stored downstream of the cerebellar cortex over the long term. In both reflexes, cerebellar LTD and LTP contribute to learning, and so do other cortical mechanisms, some of which require the inhibitory interneurons in the molecular layer. In both systems, there may be some component of memory that is initially encoded in the vestibular nuclei/DCN, and there is also clearly a component that is encoded in the cerebellar cortex. In both systems, there is evidence for local consolidation in the cortex, and there is evidence for memory transfer.

There is at least one other system that fits this general pattern: the optokinetic response. This reflex moves the eyes when the image of the visual surround "slips" on the retina. It can be triggered by whole body motion, but also by artificial situations, such as watching a train go by. Like the VOR, the optokinetic response stabilizes gaze. The optokinetic response and the VOR share very similar mechanisms of learning and memory storage.

In the remainder of this chapter, I will examine learning in two more oculomotor systems, saccades and smooth pursuit. It is not known whether they use the same mechanisms for motor learning that we have established for the VOR and NMR. It seems likely that they do. Unfortunately, mice do not make saccades or pursuit eye movements, because mice do not have a fovea. Without the mouse model it is difficult to sort out cellular mechanisms. Accordingly, I will broaden the focus a bit and ask some different questions about how we use motor learning to adjust more complex movements.

---

[3]Linear filters are selective for particular frequencies, and cause delays as a result of their frequency selectivity. Changing the gain of an adaptive filter will alter the delay in a frequency-dependent manner.

## ADJUSTING AN ORIENTING MOVEMENT

The quick, precision eye movements that we use to point the foveae of our eyes at objects of interest are known as saccades. We do not notice it, but we make about three saccades per second while scanning our surroundings. Saccades belong to the general class of movements that we might call "orienting movements"; they direct the head and body toward an interesting target. Saccades have more degrees of freedom than the VOR; each saccade has an ideal starting time, stopping time, direction, speed, and amplitude. The saccadic system is much more extensive than the VOR. Its components are distributed across all of the major subdivisions of the brain (Figure 6.2).

Saccades are very quick, and we are effectively blind while they are happening. There is no time for visual feedback anyway. This means that saccades have to be intrinsically accurate, and this is where motor learning comes in. In the lab, learning can be induced if during each saccade, the visual target of the saccade is shifted so that the saccades either consistently fall short, or consistently overshoot. Or, learning can be induced by surgically weakening the extraocular muscles of one eye, and covering the other eye with a patch.

The cerebellum, in particular lobules VI and VII of the vermis, is required for saccade motor learning (Optican & Robinson 1980). The associated deep cerebellar nucleus—the fastigial nucleus—is not absolutely required for learning. Reversible inactivation of the fastigial nucleus prevents the expression of saccade motor learning, but does not prevent some learning from occurring (Robinson *et al.* 2002). P-cells in lobules

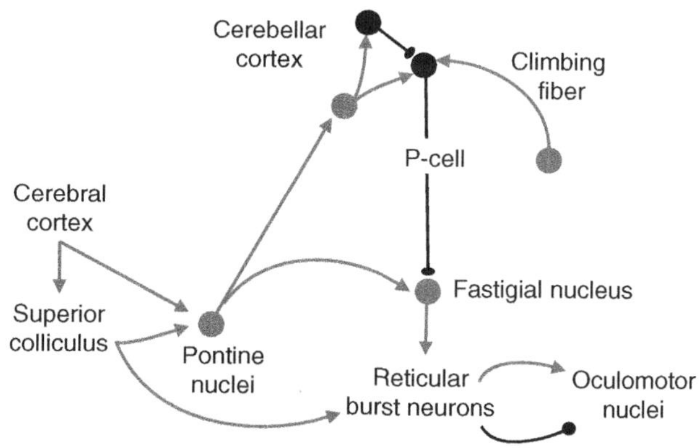


**Figure 6.2** A simplified version of the pathway for generating saccades. The motor command for saccades arises from both the cerebral cortex and the superior colliculus. Mossy fibers from the brainstem, including the pontine nuclei, ascend to the cerebellar vermis and send collaterals to the fastigial nucleus. The fastigial nucleus projects to premotor neurons that can be either excitatory or inhibitory.

VI and VII respond in a variety of ways during saccades; some generate a burst of spikes and some pause their firing. As a population, the P-cells change their responses during learning in an appropriate fashion to generate the learned change in saccade amplitude (Kojima *et al.* 2010). Together, these results point toward vermal lobules VI and VII as the normal site of memory encoding for the saccadic system.

Surprisingly, aspiration of the vermal cerebellar cortex that does not damage the fastigial nucleus causes only a temporary deficit in learning. Two months after the lesion, learning appears to be normal again (Takagi *et al.* 1998). This suggests that there are other important sites for saccade motor learning; these could include other regions of the cortex as well as the fastigial nucleus. It is also possible that the cerebral cortex, which participates in generating the command for saccades, participates in saccade motor learning.

After several weeks of saccade learning, there is a positive correlation between saccadic gain and fastigial neuronal discharges on the side away from the saccade direction (Scudder & McGee 2003). This makes sense, because the fastigial nucleus sends a crossed excitatory projection to premotor neurons that excite motoneurons during saccades. For example, a neuron in the right fastigial nucleus would give an added push to leftward saccades with its excitatory burst. Increasing the burst may, therefore, generate a larger movement. Meanwhile, neurons in the fastigial nucleus on the other side (left, in our example) also respond, and during the saccade they activate a group of brainstem neurons that *inhibit* motoneurons during leftward saccades.[4] To increase the size of the saccade, these discharges must become smaller, and they do (Inaba *et al.* 2003).

What all this means is that motor learning adjusts the responses of fastigial neurons on the left and right sides of the cerebellum in opposite directions. It appears that in both the VOR and saccadic systems, changes in gain are linked to quantitative adjustments in vestibular or DCN neuronal responses.

In addition to saccade speed, timing is an important variable in saccade learning. Some fastigial neurons fire to terminate saccades toward the side they are on; the fastigial burst can act as a brake. After several weeks of learning a larger saccade size, the burst becomes delayed. Learning a smaller size results in an earlier burst (Scudder & McGee 2003). This is what we would expect, because when we learn to increase the saccade size, the saccade must end a bit later. Meanwhile, upstream of the fastigial neurons, as the duration of the saccade increases more P-cells begin to respond late in the saccade, possibly extending it by delaying the

[4]These inhibitory neurons are the "inhibitory burst neurons" (IBNs), which are located in the medullary reticular formation. The function of IBNs may be to relax the antagonist muscles during the saccade.

fastigial burst (Catz *et al.* 2008). This timing mechanism takes advantage of rebound firing by fastigial nuclear neurons, to generate a braking burst at precisely the right time.

There is some evidence that saccadic memory consolidates after several days of training. Although the learning mechanism can saturate during extended sessions, further learning again becomes possible after the memory is fully consolidated. This is consistent with separate sites for storage of short- and long-term memory in the saccadic system (Robinson *et al.* 2006).

### *Specificity of saccade motor learning*

If error signals guide saccadic motor learning via cerebellar LTD, we might expect climbing-fiber discharge rates to increase right after a saccade fails to reach its target, and they do. The opposite pattern is seen if the saccade overshoots the target (Soetedjo & Fuchs 2006). The relationship between these errors and complex spikes may not be consistent, because in at least some cases, complex spikes cause a decrease in the size of saccades while a pause in the complex spikes causes an increase (Catz *et al.* 2008). But let's assume that for saccadic motor learning, as for VOR motor learning, the direction of the error is encoded by complex spikes. This means that P-cells may undergo LTD of their excitatory inputs, which could disinhibit the fastigial nuclei in situations where they need to fire at a higher rate; and LTP could occur in the opposite situation. These changes could increase the intensity of the burst that accelerates the saccade and/or delay the burst that stops it. On the cellular level, learning in the saccadic system may operate much like learning in the VOR.

Marr originally proposed that different P-cells would participate in learning particular movements. During saccadic learning, each climbing fiber responds selectively to a particular direction and size of error (Soetedjo *et al.* 2008). Because each P-cell receive input from only one climbing fiber, this means that in the saccadic system different populations of P-cells must participate in learning of changes in saccade gain for saccades of specific directions, as Marr's hypothesis implies (Chapter 2). This would explain why learning does not transfer well to saccades in different directions, during which other populations of P-cells would be active (Noto *et al.* 1999).

## ADJUSTING A TRACKING MOVEMENT

The "smooth pursuit" system moves the eyes to track a visual target that is moving slowly enough (if the eyes lag too far behind the target, saccades are used). By matching eye speed to the speed of a moving object, pursuit keeps the image of the object on the fovea, so we can see it clearly. The pursuit system automatically makes use of visual feedback to adjust eye speed. Its neural control system includes most of the elements of the networks in Figures 6.1 and 6.2; both the ventral paraflocculus and the vermis are involved in pursuit.

To hit a tennis ball that is going to bounce, we have to make a prediction about the speed and direction of the ball after the bounce. Pursuit has a similar problem because visual feedback takes roughly 100 ms. During that time lag, the target can change direction, but the eyes are moving at a predetermined speed. However, if the change is predictable, then the pursuit gain can be adjusted preemptively, keeping the image near the fovea. The pursuit system can learn to make these adjustments for targets that reliably accelerate or decelerate.

It is important to notice that in the saccadic and pursuit control systems, where the cerebral cortex is involved in generating the motor command, the cerebellar cortex still is responsible for learning fine motor adjustments. Lesions of the flocculus reduce the gain of tracking, and this deficit is not corrected over time. As we saw for saccadic learning, pursuit learning is selective for the particular direction (say, rightward) that the target was moving in during training, and this selectivity depends on the direction that the eyes are moving in, not the direction of image motion (Kahlon & Lisberger 1996). This is consistent with learning in the cerebellar cortex itself, where movements are already encoded in their proper motor coordinates.[5]

In the ventral paraflocculus, P-cells carry eye-velocity and eye-acceleration signals during pursuit. The same P-cells also show changes in their discharge patterns during pursuit learning (i.e., in their on-directions). Complex spikes respond to motion of the target image on the retina, and encode the direction of the retinal error during pursuit learning, suggesting that learning may be guided by climbing fibers (Kahlon & Lisberger 2000). If cerebellar LTD occurs during motor learning, the occurrence of a complex spike during one trial should cause a depression of simple-spike response on the next trial. Similarly, with cerebellar LTP, a pause in the complex spike discharges should cause an increase in simple-spike response on the next trial. Furthermore, if these changes are responsible for learning, they should accompany changes in behavior, consistent with the response properties of the P-cells. All three of these predictions turn out to be correct (Medina & Lisberger 2008). It seems likely that floccular P-cells cause the change in oculomotor behavior during learning, simply by exaggerating responses that they already exhibit during smooth pursuit. This would be consistent with Marr's hypothesis.

Lesions of the oculomotor vermis (lobules VI and VII) also impair pursuit learning (Takagi *et al.* 2000). Strangely, the incidence of complex

[5]Motor coordinates and sensory coordinates are not the same, because the coordinate system depends on the sensing or effecting organ and its geometry.

spikes in the vermis actually can increase during learning (Dash *et al.* 2010). The buildup of complex spikes might reflect a process of integration of performance errors (Catz *et al.* 2008). This idea does not fit into the dogma of Marr and Albus. We have seen that multiple mechanisms participate in eyeblink and VOR motor learning. Various mechanisms may also contribute to pursuit learning, depending on the context. It is also possible that the vermis and the flocculus perform fundamentally different functions during pursuit learning.

## CONCLUSIONS

The cerebellum uses multiple cellular mechanisms, including but not restricted to LTD, to correct the amplitude and speed of eye movements. Memory of these improvements is stored, in part, in the DCN and vestibular nuclei. Some new memory is also stored in the cerebellar cortex. In broad outline, learning appears to be similar across at least some oculomotor systems (the VOR, saccades, and smooth pursuit). But at the same time, different oculomotor systems impose different requirements on the cerebellum. Some movements require timing of delay intervals, and others require very precise adjustments of feed-forward gain. Because of its multitude of sites and mechanisms for plasticity, the cerebellum has the flexibility to meet the needs of our diverse motor systems.

## REFERENCES


Anzai, M., Kitazawa, H. & Nagao, S. (2010) Effects of reversible pharmacological shutdown of cerebellar flocculus on the memory of long-term horizontal vestibulo-ocular reflex adaptation in monkeys. *Neurosci. Res.*, **68**, 191–198.

Boyden, E.S. & Raymond, J.L. (2003) Active reversal of motor memories reveals rules governing memory encoding. *Neuron.*, **39**, 1031–1042.

Boyden, E.S., Katoh, A., Pyle, J.L., Chatila, T.A., Tsien, R.Y. & Raymond, J.L. (2006) Selective engagement of plasticity mechanisms for motor memory storage. *Neuron.*, **51**, 823–834.

Broussard, D.M., Bronte-Stewart, H.M. & Lisberger, S.G. (1992) Expression of motor learning in the response of the primate vestibuloocular reflex pathway to electrical stimulation. *J. Neurophysiol.*, **67**, 1493–1507.

Catz, N., Dicke, P.W. & Thier, P. (2008) Cerebellar-dependent motor learning is based on pruning a Purkinje cell population response. *Proc. Natl. Acad. Sci. USA*, **105**, 7309–7314.

Dash, S., Catz, N., Dicke, P.W. & Thier, P. (2010) Specific vermal complex spike responses build up during the course of smooth-pursuit adaptation, paralleling the decrease of performance error. *Exp. Brain Res.*, **205**, 41–55.

Gao, Z., Van Beugen, B.J. & De Zeeuw, C.I. (2012) Distributed synergistic plasticity and cerebellar learning. *Nat. Rev. Neurosci.*, **13**, 619–635.
Inaba, N., Iwamoto, Y. & Yoshida, K. (2003) Changes in cerebellar fastigial burst activity related to saccadic gain adaptation in the monkey. *Neurosci. Res.*, **46**, 359–368.
Ito, M. (1972) Neural design of the cerebellar motor control system. *Brain Res.*, **40**, 81–84.
Kahlon, M. & Lisberger, S.G. (1996) Coordinate system for learning in the smooth pursuit eye movements of monkeys. *J. Neurosci.*, **16**, 7270–7283.
Kahlon, M. & Lisberger, S.G. (2000) Changes in the responses of Purkinje cells in the floccular complex of monkeys after motor learning in smooth pursuit eye movements. *J. Neurophysiol.*, **84**, 2945–2960.
Kassardjian, C.D., Tan, Y.F., Chung, J.Y., Heskin, R., Peterson, M.J. & Broussard, D.M. (2005) The site of a motor memory shifts with consolidation. *J. Neurosci.*, **25**, 7979–7985.
Ke, M.C., Guo, C.C. & Raymond, J.L. (2009) Elimination of climbing fiber instructive signals during motor learning. *Nat. Neurosci.*, **12**, 1171–1179.
Kim, C.H., Oh, S.H., Lee, J.H., Chang, S.O., Kim, J. & Kim, S.J. (2012) Lobule-specific membrane excitability of cerebellar Purkinje cells. *J. Physiol.*, **590**, 273–288.
Kimpo, R., Boyden, E.S., Katoh, A., Ke, M.C. & Raymond, J.L. (2005) Distinct patterns of stimulus generalization of increases and decreases in VOR gain. *J. Neurophysiol.*, **94**, 3092–3100.
Kojima, Y., Soetedjo, R. & Fuchs, A.F. (2010) Changes in simple spike activity of some Purkinje cells in the oculomotor vermis during saccade adaptation are appropriate to participate in motor learning. *J. Neurosci.*, **30**, 3715–3727.
Lisberger, S.G., Pavelko, T.A., Bronte-Stewart, H.M. & Stone, L.S. (1994a) Neural basis for motor learning in the vestibulo-ocular reflex of primates. II. Changes in the responses of horizontal gaze velocity Purkinje cells in the cerebellar flocculus and ventral paraflocculus. *J. Neurophysiol.*, **72**, 954–973.
Lisberger, S.G., Pavelko, T.A. & Broussard, D.M. (1994b) Neural basis for motor learning in the vestibulo-ocular reflex of primates. I. Changes in the responses of brain stem neurons. *J. Neurophysiol.*, **72**, 928–953.
Luebke, A.E. & Robinson, D.A. (1994) Gain changes of the cat's vestibulo-ocular reflex after flocculus deactivation. *Exp. Brain Res.*, **98**, 379–390.
Maekawa, K. & Simpson, J.I. (1973) Climbing fiber responses evoked in vestibulocerebellum of rabbit from visual system. *J. Neurophysiol.*, **36**, 649–666.
McElligott, J.G., Beeton, P. & Polk, J. (1998) Effect of cerebellar inactivation by lidocaine microdialysis on the vestibuloocular reflex in goldfish. *J. Neurophysiol.*, **79**, 1286–1294.
McElvain, L.E., Bagnall, M.W., Sakatos, A. & du Lac, S. (2010) Bidirectional plasticity gated by hyperpolarization controls the gain of postsynaptic firing responses at central vestibular nerve synapses. *Neuron.*, **68**, 763–775.
Medina, J.F. & Lisberger, S.G. (2008) Links from complex spikes to local plasticity and motor learning in the cerebellum of awake-behaving monkeys. *Nat. Neurosci.*, **11**, 1185–1191.

Nagao, S. (1983) Effects of vestibulocerebellar lesions upon dynamic characteristics and adaptation of vestibulo-ocular and optokinetic responses in pigmented rabbits. *Exp. Brain Res.*, **53**, 152–168.

Nelson, A.B., Gittis, A.H. & du Lac, S. (2005) Decreases in CaMKII activity trigger persistent potentiation of intrinsic excitability in spontaneously firing vestibular nucleus neurons. *Neuron*, **46**, 623–631.

Noto, C.T., Watanabe, S. & Fuchs, A.F. (1999) Characteristics of simian adaptation fields produced by behavioral changes in saccade size and direction. *J. Neurophysiol.*, **81**, 2798–2813.

Optican, L.M. & Robinson, D.A. (1980) Cerebellar-dependent adaptive control of primate saccadic system. *J. Neurophysiol.*, **44**, 1058–1076.

Pastor, A.M., De La Cruz, R.R. & Baker, R. (1994) Cerebellar role in adaptation of goldfish vestibuloocular reflex. *J. Neurophysiol.*, **72**, 1383–1394.

Rambold, H., Churchland, A., Selig, Y., Jasmin, L. & Lisberger, S.G. (2002) Partial ablations of the flocculus and ventral paraflocculus in monkeys cause linked deficits in smooth pursuit eye movements and adaptive modification of the VOR. *J. Neurophysiol.*, **87**, 912–924.

Raymond, J.L. & Lisberger, S.G. (1998) Neural learning rules for the vestibulo-ocular reflex. *J. Neurosci.*, **18**, 9112–9129.

Robinson, F.R., Fuchs, A.F. & Noto, C.T. (2002) Cerebellar influences on saccade plasticity. *Ann. N. Y. Acad. Sci*, **956**, 155–163.

Robinson, F.R., Soedetjo, R. & Noto, C. (2006) Distinct short-term and long-term adaptation to reduce saccade size in monkey. *J. Neurophysiol.*, **96**, 1030–1041.

Schonewille, M., Belmeguenai, A., Koekkoek, S.K.E., Houtman, S.H., Boele, H.J., van Beugen, B.J., Gao, Z., Badura, A., Ohtsuki, G., Amerika, W.E., Hosy, E., Hoebeek, F.E., Elgersma, Y., Hansel, C. & De Zeeuw, C.I. (2010) Purkinje cell-specific knockout of the protein phosphatase PP2B impairs potentiation and cerebellar motor learning. *Neuron.*, **67**, 618–628.

Schonewille, M., Gao, Z., Boele, H.J., Veloz, M.F.V., Amerika, W.E., Simek, A.A.M., De Jeu, M.T.G., Steinberg, J.P., Takamiya, K., Hoebeek, F.E., Linden, D.J., Huganir, R.L. & De Zeeuw, C.I. (2011) Reevaluating the role of LTD in cerebellar motor learning. *Neuron.*, **70**, 43–50.

Scudder, C.A. & McGee, D.M. (2003) Adaptive modification of saccade size produces correlated changes in the discharges of fastigial nucleus neurons. *J. Neurophysiol.*, **90**, 1011–1026.

Soetedjo, R. & Fuchs, A.F. (2006) Complex spike activity of Purkinje cells in the oculomotor vermis during behavioral adaptation of monkey saccades. *J. Neurosci.*, **26**, 7741–7755.

Soetedjo, R., Kojima, Y. & Fuchs, A.F. (2008) Complex spike activity in the oculomotor vermis of the cerebellum: a vectorial error signal for saccade motor learning?. *J. Neurophysiol.*, **100**, 1949–1966.

Takagi, M., Zee, D.S. & Tamargo, R.J. (1998) Effects of lesions of the oculomotor vermis on eye movements in primate: saccades. *J. Neurophysiol.*, **80**, 1911–1931.

Takagi, M., Zee, D.S. & Tamargo, R.J. (2000) Effects of lesions of the oculomotor cerebellar vermis on eye movements in primate: smooth pursuit. *J. Neurophysiol.*, **83**, 2047–2062.

Titley, H.K., Heskin-Sweezie, R., Chung, J.Y., Kassardjian, C.D., Razik, F. & Broussard, D.M. (2007) Rapid consolidation of motor memory in the vestibulo-ocular reflex. *J. Neurophysiol.*, **98**, 3809–3812.

Titley, H.K., Heskin-Sweezie, R. & Broussard, D.M. (2010) The bidirectionality of motor learning in the vestibulo-ocular reflex is a function of cerebellar mGluR1 receptors. *J. Neurophysiol.*, **104**, 3657–3666.

Wulff, P., Schonewille, M., Renzi, M., Viltono, L., Sassoe-Pognetto, M., Badura, A., Gao, Z., Hoebeek, F.E., van Dorp, S., Wisden, W., Farrant, M. & De Zeeuw, C.I. (2009) Synaptic inhibition of Purkinje cells mediates consolidation of vestibulo-cerebellar motor learning. *Nat. Neurosci.*, **12**, 1042–1049.

## FURTHER READING

Raymond, J.L., Lisberger, S.G. & Mauk, M.D. (1996) The cerebellum: a neuronal learning machine? *Science*, **272**, 1126–1131.

Boyden, E.S., Katoh, A. & Raymond, J.L. (2004) Cerebellum-dependent learning: the role of multiple plasticity mechanisms. *Ann. Rev. Neurosci.*, **27**, 581–609.

Hopp, J.J. & Fuchs, A.F. (2004) The characteristics and neuronal substrate of saccadic eye movement plasticity. *Prog. Neurobiol.*, **72**, 27–53.

# 7 Perfecting Limb Movements by Motor Learning

Reaching, throwing, manipulating objects, and even walking all need to be learned in order to be effective. This is because we have to deal with the built-in time delays that are associated with sensory feedback, which we need for guiding movements. As this is being written, Erik Weihenmayer—who became blind at the age of 14—is preparing to navigate the Grand Canyon *in a kayak*. Surprisingly, Weihenmayer comments that "without the eyes... it's like sensory overload" (Olsen 2012). Weihenmayer's description seems counterintuitive because if visual information is present, we pay a lot of attention to it, especially in a situation like whitewater kayaking. But in fact, somatosensory and kinesthetic inputs are crucial for kayaking, skiing, surfing, and similar sports (while evidently, vision is not). The cerebellum has access to many different kinds of information, and the information is encoded in a variety of coordinate systems. Somehow, we must learn to assemble them into a single system that can guide movement.

After making a mistake in a limb movement—indeed, in any kind of movement—it can take 100 ms or more to find out about the mistake from sensory feedback, and it is already too late to react. Because of the delay problem we try to plan movements, taking into account potential problems (e.g., learning to accelerate or decelerate the eyes during smooth pursuit). This planning is called "predictive control." In patients with cerebellar damage, predictive control is impaired much more than reactive control is (Bastian 2006).

We constantly work to perfect our limb movements, but we do not always do it on a conscious level. Introspection gives very little insight into how the learning (or, indeed, predictive control itself) is accomplished. How can a multi-jointed limb reach toward an object? How do we coordinate the muscles? How are errors used, and do we learn in other ways? How is the cerebellum involved in learning these motor skills?

One way to generate appropriate predictive command signals for the muscles is to use internal models. For example, a "forward" internal

*The Cerebellum: Learning Movement, Language, and Social Skills*, First Edition. Dianne M. Broussard.

model represents the physical characteristics of the arm. The model can predict what sensory input should result from the movement (e.g., touching the target object). The model can be fine-tuned by calculating the error between what it predicts and what actually happens. The error signal is used to teach the model (Wolpert *et al.* 1998). We use two kinds of internal models of our limbs: Dynamic models and kinematic models. Dynamic models predict the time course of the movement. In order to do this, dynamic models must take into account properties like inertia, friction, and elasticity of the components of the arm. These factors all affect the time course of the movement. Dynamic models also use the intrinsic coordinate system of the limb. In the intrinsic coordinate system, each axis of rotation for each joint must be assigned its own coordinate. Dynamic models are capable of predicting what the limb might be expected to do with a given command signal, based on its properties. Dynamic internal models are really essential for all kinds of movement. They probably evolved relatively early, and some of them are implemented by older parts of the brain.

Kinematic models represent the desired change in position in Cartesian coordinates (the "vector in space"). Dynamic models are thought to be calibrated using sensory feedback from the limb itself, but kinematic models can use any sensory information, and are useful for prey capture, for example. Kinematic models can learn new gains, for example, in response to visual errors (Krakauer *et al.* 1999). One possibility is that we use our excellent motion perception and depth vision to update our kinematic models, as a cross-check on the older system of dynamic models.

## UPDATING DYNAMIC MODELS

The sense of touch and proprioception[1] provide error signals to guide the process of updating dynamic models for limb movements. If motor learning for limb movements is done in the cerebellum, both the model itself and the errors should be represented in the cerebellum, and this seems to be the case. Sensory input from the forelimbs reaches both the anterior and the posterior lobes of the cerebellum. Purkinje cells (P-cells) in the paravermal regions of lobules IV and V respond to somatosensory inputs from the forelimbs and hands or paws, as do neurons in the nucleus interpositus.

The nucleus interpositus is a site for motor memory for limb movements, suggesting that it may also be a storage site for the model. The interpositus participates in classically conditioned withdrawal responses of

[1]Proprioceptive signals arise from the joints and muscles and report the position and dynamic rotation of each joint.

the limbs, in some details of reaching and grasping movements, in flexing the paw, and in precision paw placement, but not in unconditioned withdrawal reflexes (Bracha *et al.* 1999). In monkeys, P-cells in the vermis and paravermis of lobules V and VI, and neurons in the DCN (the dentate, interpositus, and fastigial nuclei), respond during arm movements (MacKay 1988). Stimulation of lobule V of the anterior lobe leads to movements of the arm or forelimb in most studies, demonstrating that the signals in the anterior lobe are actually motor commands, at least in part.

In the posterior lobe, the situation is more complicated; the motor representation of the arm is the caudal part of in lobule VII, while at least part of the sensory representation is further rostral in crus II (Manni & Petrosini 2004). Nevertheless, complex spikes do encode errors in arm movements. In one famous study, monkeys learned to correct for perturbing forces that were imposed by a torque motor. The monkeys had to cancel the perturbation perfectly and maintain hand and arm position. This required predictive control. Complex spikes initially responded to the external forces. However, as learning progressed, the complex-spike response got smaller (Gilbert & Thach 1977). Decreases in simple-spike activity were associated with the decrease in complex spikes, consistent with LTD as a learning mechanism. Figure 7.1 shows part of the circuitry underlying motor learning for arm movements in the posterior lobe and the dentate nucleus.

The error signals carried by complex spikes sometimes originate from the sense of touch. The anterior lobe and the posterior lobe share

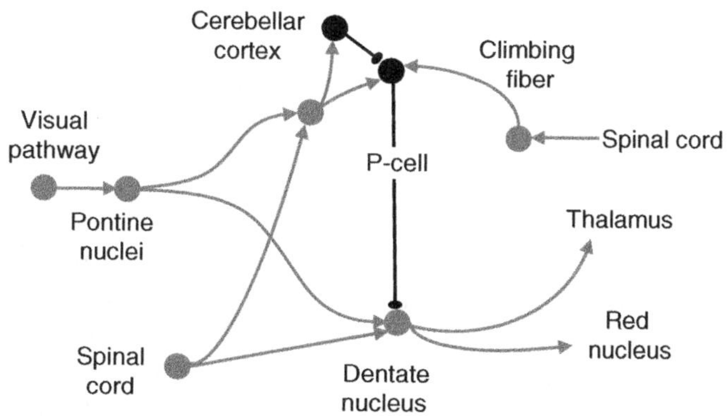


**Figure 7.1** A simplified version of the circuitry for learning of visually guided reaching by the posterior lobe. The visual cortex, association cortex and spinal cord provide some of the sensory input guiding the reach. Other inputs guiding the reach may include oculomotor signals. Error feedback is provided by proprioceptors and somatosensory receptors in the hand and arm, and also by visual signals (not shown). Both influence climbing fiber signals. The output of the dentate nucleus reaches motoneurons via the red nucleus and also reaches motor cortex by way of the thalamus.

responsibility for arm movements, depending on the context, and the error signals for goal-directed reaching and throwing arm movements are represented in the hemispheres, not the vermis. Responses can be measured in the lateral parts of lobules V and VI, in a region associated with somatosensory and motor representation of the hand. In humans that are learning a reaching task, either the presence of an unexpected somatosensory stimulus or the absence of an expected one can cause increased activity (Schlerf *et al.* 2012). All of this seems to mean that the cerebellum is involved in learning accurate reaching, using feedback from the arm.

The cerebellum is also involved in another kind of learning called a force-field task, in which practice with a novel mechanical system updates the dynamic model of the arm, similar to learning to shift gears in a car (Maschke *et al.* 2004). The error signal for the force-field task is somatosensory and proprioceptive. Dynamic models of the arm could be updated directly by complex-spike error signals in the force-field tasks, as we think they are in reaching. But it seems more likely that in the force-field task, the memory is transferred from the motor cortex to the cortex of the cerebellar anterior lobe, during rapid consolidation of the memory in the hours following practice sessions (Shadmehr & Holcomb 1997).

Single-unit recordings from P-cells in monkeys that have learned the force-field task have failed to find evidence that the memory remains in the cerebellar cortex for the long term (Pasalar *et al.* 2006). Perhaps the deep nuclei store the memory for the long term. But why would the memory need to be shifted twice? One possibility is that the cerebellum is creating a working model of the cerebral cortex (Ito 1993). The idea here is that the cerebellum might mimic the function of the cortex as it integrates sensory information and generates a motor command. The working model could then perform the same integration and command generation, and carry it out more rapidly and efficiently. The cerebral cortex, having delegated most of the work, could move on to the next problem. Delegation to the cerebellum is also a possibility for reaching tasks and for many other cerebral cortical tasks.

The linkage of the cerebellum to forebrain processing centers is not only necessary for long-term memory, but may actually act as another site for memory storage in the synapses that make up the linkage. Rats, like humans, can learn to compensate for mechanical perturbations that are analogous to the force-field task. The learning requires several weeks, suggesting the measured responses included both new and consolidated memory. Interestingly, after the task had been thoroughly learned, cerebellothalamic synapses in these rats showed an increase in the proportion of docked synaptic vesicles (Aumann & Horne 1999). This change, strategically located between the cerebellum and the forebrain, suggests that the connection of the cerebellum with other parts of the brain participates in long-term memory storage.

## THROWING AND POINTING

Any visually guided movement can be corrected by a visual error signal. Throwing is a good example. In the nineteenth century, Hermann von Helmholtz knew that wearing prism goggles over the eyes causes systematic deviations in throwing. But the errors disappear within a few trials, suggesting that we learn to compensate for the inaccuracy, using the visual feedback at the end of each throw. Even more interesting, if the prisms are removed after a few days of practice with the prisms, errors begin to occur again. These errors mean that a persistent change has been made.

The prism-learning situation is complex, because the visual error (say, landing consistently to the right of the bull's eye) and the movement itself must be represented in completely different coordinate systems. Somehow, the different signals have to be compared.

The cerebellar posterior lobe seems to manage this complex task. The error signal may be used to update a kinematic internal model, in spatial coordinates.[2] This could happen in lobule VI of the posterior lobe, where input from the arms converges with auditory and visual inputs. Multisensory convergence may be useful for updating kinematic models in general.

Damage to the anterior lobe causes abnormalities of voluntary movement in general (ataxia), but does not always cause deficits in prism learning. Meanwhile, subjects with damage to the posterior-lobe hemispheres cannot learn accurate throwing with prisms, but they can use the affected arm for other coordinated tasks such as climbing (Baizer *et al.* 1999).

Lesions of the inferior olive also prevent any prism learning, indicating that the cerebellum is the site of memory encoding (Martin *et al.* 1996). Both the normal guidance of the movement by vision and the visual error signal during learning employ climbing-fiber signals. Accordingly, complex spikes in the hemisphere of lobules IV–VI actually encode two kinds of visuomotor "errors": the location of the goal, which causes a discharge early in the task, and the artificial movement error (Kitazawa *et al.* 1998).

Pointing movements, including those that we make using a computer mouse or joystick, can also be guided (and corrected) by vision. Visually simulated rotation errors while subjects are learning to move a cursor toward a target cause more activation than the movement alone does. The activity increases in lobules V, VI, and VIII. If the error is interpreted as being due to a faulty motor command, instead of simply due to motion of the target, then learning occurs (Diedrichsen *et al.* 2005). As the subjects learn to compensate for the artificial rotation, activity increases over a

[2]Spatial coordinates are the coordinates with respect to the outside world, or "space."

wide region of the hemisphere. Here is the interesting part: the activated region becomes smaller and more focused as learning progresses. This could reflect changes in complex-spike activity, but the changes seem too large for that. It could mean instead that a particular internal model, represented by activity in a particular locus, has been selected during learning. In this variant of Marr's hypothesis, multiple kinematic models coexist, and all receive copies of the error signal, leading to the selection of one that is appropriate for each task (Imamizu *et al.* 2000).

Do visual and somatosensory error signals interact with each other in the cerebellum? The simulated visual error signals in the pointing task activate some of the same cerebellar regions that are active in the force-field task (Diedrichsen *et al.* 2005). Specifically, both learning situations activate crura I and II (Donchin *et al.* 2012). This is interesting, because visual and proprioceptive errors are encoded in different coordinate systems. We can speculate that the intrinsic cerebellar circuitry might be involved in conversions among coordinate systems for the different kinds of sensory input, which can be computationally intensive. This idea does not seem to fit with the concept that errors arrive, precalculated, on climbing fibers. However, feedback to the olive could be involved in learning the conversion.

Force-field tasks may update a dynamic model, while pointing seems likely to use a kinematic model. The overlapping activation may be involved in comparing the predictions of kinematic and dynamic models. But we would also expect some separation of activation in the two tasks. In fact, a small region on the border of lobules V and VI is selectively activated by visual errors (Diedrichsen *et al.* 2005) while another region in the anterior lobe is selectively activated by somatosensory ones (Donchin *et al.* 2012). These separate error signals could represent different internal models (kinematic and dynamic) that are updated by the different errors.

There is another problem with learning to throw with prisms: feedback of the error is restricted to the end of the task, when the subject realizes that he has missed the target. This suggests that cognitive feedback (i.e., thinking about the error) could be used in this task. The posterior lobe has been proposed to participate in several other cognitive functions, which I will describe later on.

What about the long term? Memory for skilled arm movements does consolidate, and consolidation may involve memory transfer (Shadmehr & Brashers-Krug 1997). Learning of accurate reaching has fast and slow components, suggesting multiple mechanisms for memory encoding (Shadmehr *et al.* 2010). Recent results with cerebellar patients were consistent with the idea that rapid learning is selectively lost in cerebellar damage (Tseng *et al.* 2007). This could reflect redistribution of memory after it is encoded. In a task with simulated visual errors, strokes affecting lobule VI were likely to affect reversal, while strokes affecting lobule V were more likely to affect retention of the memory (after 1–7 days) (Werner *et al.* 2010).

## SEQUENCE LEARNING

Skilled movements of the fingers, like accurate throwing, require some practice to execute well. Playing a musical instrument, for example, almost certainly requires cerebellar motor learning. In the lab, finger-tapping—a form of sequence learning in which keys are depressed rapidly in a learned sequence—stands in for some aspects of musical performance. The cerebellum, including the hemispheres, the vermis (both anterior and posterior), and the deep nuclei, is definitely more active during learning of a new finger-tapping task (using auditory feedback) than it is during performance of the same task after learning it (Jenkins *et al.* 1994). Very broad regions of the cerebellar cortex, including lobules III, IV, V, VI, and IX of the vermis and VI (crus I) and VIII of the hemispheres, became active during this kind of sequence learning (Doyon *et al.* 2002; Penhune & Doyon 2002). Clearly, both the anterior and posterior lobes of the cerebellar cortex are involved in sequence learning.

We are generally aware of movement sequences, and exert some cognitive control over them, but this is not always necessary. Some sequences may be learned without conscious awareness that we are learning a sequence. The cognitive and motor components of sequence learning are difficult to separate, but there have been some attempts. One approach is to have the subjects explicitly memorize the sequence before practicing. The idea behind this—which may or may not be correct—is that the memorization contains the "cognitive component." When memorization is done in advance, the lobule V vermis and hemispheres are still highly active during the actual practice sessions, more so than they are during random finger-tapping (Penhune & Doyon 2002).

So far, research seems to support the idea that the cerebellar anterior lobe may be required for learning some aspects of motor coordination (and possibly for speeding up performance). During the learning of a sequence of visuomotor responses, the cerebellum is necessary for the shortening of reaction times that normally occurs once the sequence has been learned (Molinari *et al.* 1997). The posterior lobe, on the other hand, may support some cognitive components of sequence learning. For example, it could be involved in learning the sequence itself. It could learn a context for each movement in the sequence, based on sensory input. For example, hearing each note or phrase in a musical piece that one is playing may access the memory of the next note or phrase. Or, the cerebellum could just execute previously-learned sequences with accurate timing.

As proposed earlier, the cerebellum may create working models of cortical functions. If the cerebellum provides motor control independently, in parallel with the forebrain areas—such as frontal lobes—that are responsible for movement planning, it may take control once it has learned a good model, and could execute the sequence with improved speed.

Although practice is certainly required, improvement in performance during periods where no practice occurs is a very common experience for musicians. The improvement could reflect the transfer of memory from forebrain to cerebellar networks—or the other way around. For example, learning a sequence of long and short depressions of the same key activates the cerebellar cortex early in the learning process, but not later in the process (Penhune & Doyon 2002). It has been proposed that "fast" learning during a single session may be exclusively cerebellar, and may be followed by transfer over the next several hours to days (Doyon *et al.* 2003). Of course, this task is much less complex than actual keyboard playing, where multiple keys would be involved. In learning to press a sequence of different keys, the dentate and interpositus nuclei become activated after the task is fairly well learned (Doyon *et al.* 2002).

Muscimol injections into the dorsal or central dentate nucleus impair performance of button-pressing sequences that have been memorized and practiced for several months. This means that the dentate nucleus is one long-term storage site for button-pressing sequences (Lu *et al.* 1998). Since the deficit is limited to the hand on the same side as the injection, and the primary motor cortex projects to motoneurons for only one hand, the downstream connections of the dentate nucleus with primary motor cortex may also be involved in storing the memory. The cerebellar cortex may also be a site for memory storage.

It is interesting that while for the force-field task, the cerebellar anterior lobe (and DCN) are sites for long-term storage, the posterior lobe is more important for sequence learning. The two tasks are quite different. Unlike force-field learning, sequence learning does not require the updating of internal models of the limb. However, it does require storage of information relating the movement spatially to the outside world (specifically, the buttons). It also includes a cognitive component: we need to learn the sequence in order to perform the task, although it may become automatic at some point.

The role of internal models has not been addressed yet for sequence learning, but sequence might be a useful component in internal models. For example, in simulations, internal models that generate commands for arm movements can make use of cues that are related to movement order in a sequence (Wainscott *et al.* 2005). The same may be true for finger-tapping. For example, the cerebellum may learn to associate sensory cues with the next movement in the sequence. Also, some parameters of internal models depend critically on the current contraction state of the muscle and speed and position of the joints, which depend in turn on the previous movement in the sequence.

Of course, sequence learning is only a small part of musical performance. Even a very detailed and precisely timed memorized sequence would not be sufficient when improvising a new melody or for musical interpretation. We are very far from a full understanding of the role of the cerebellum in music. However, this has not stopped several authors from making interesting proposals, and I will discuss some of these in Chapter 12.

## STEPPING AND CHANGES TO GAIT

Locomotion is one of the most difficult motor skills to quantify. It is also interesting because it straddles the boundary between learned and innate movement patterns. Humans must learn to walk by trial and error. However, many animals require little or no practice, and the presence of a central pattern generator in the spinal cord suggests that basic locomotor patterns might be innate. But we can learn many specialized patterns of locomotion when necessary, using sensory feedback. Updating of dynamic models using the cerebellum seems the most likely scenario for learning to hop, skip, and jump, or just to walk on an uneven surface. Foot placement and timing, balance, and coordination of different muscle groups during walking may also be cerebellar functions that we learn a bit earlier in life.

Lobules IV and V of the anterior lobe receive somatosensory and proprioceptive input from the legs and feet. The intermediate zone of these lobules (just lateral to the vermis) is connected with spinal pattern generators for locomotion, and may contribute to stepping. Another crucial function—especially for bipedal locomotion—is balance. Cerebellar patients who have damage to the fastigial nucleus and the anterior lobe are likely to show difficulties with balance during walking, while patients with damage to the nucleus interpositus and the adjacent part of the dentate nucleus are more likely to have trouble with foot placement (Ilg *et al.* 2008). The pattern seems to be that the vermis of the anterior lobe specializes in balance, and the more lateral regions function in moving the feet accurately. The posterior lobes are less important for locomotion (Martin *et al.* 1996).

Figure 7.2 shows a simplified version of the anterior lobe and associated circuitry controlling the basic rhythm of locomotion. The circuitry for balance during locomotion, which relies on the vestibular system, is not shown. The circuitry that controls foot placement, and which includes the dentate nucleus rather than the fastigial nucleus, is also not shown, and is probably similar to the circuitry for reaching in Figure 7.1.

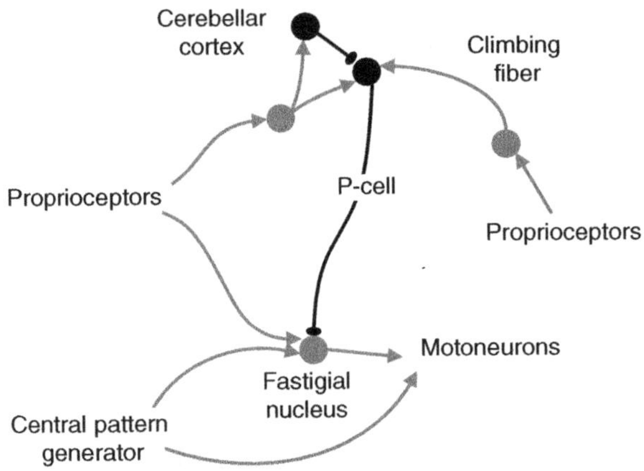


**Figure 7.2** The basic circuitry that generates and modifies the rhythm for walking. The central pattern generator in the spinal cord provides input to the fastigial nucleus, which projects in turn to the motoneurons of the spinal cord. Proprioceptors are located in the muscles and joints of the legs and trunk, and project to the dorsal root ganglia of the spinal cord. Along with proprioceptors, somatosensory receptors on the soles of the feet also contribute sensory feedback (by a similar anatomical pathway) during walking.

The split-belt treadmill, with belts that move at different speeds under each foot, generates reproducible errors in locomotion. To walk on it, subjects have to learn to match their step lengths for the "fast" and "slow" feet using predictive control. Patients with cerebellar damage often fail to match their step lengths on the split-belt. Unlike the reaching task, failure in the split-belt task is correlated with ataxia. However, the same patients that fail to learn can still make some immediate corrections to their gait, perhaps using a region of the cerebellum that has escaped damage (Morton & Bastian 2006). Gait ataxia is not always the same, and different effects are seen following damage to the medial and lateral cerebellar zones (Ilg *et al.* 2008).

While cerebellar damage prevents subjects from learning to walk on the split-belt, the cerebral cortex is sometimes unnecessary. Decerebrate[3] cats, which are lacking the entire forebrain, can learn to walk on a split-belt treadmill. During learning, climbing-fiber discharge in the vermis of lobule V becomes timed to particular phases of locomotion (Yanagihara

[3]In this archaic procedure, the animal was anesthetized and the brainstem was transected or lesioned just rostral to the superior colliculus, isolating the higher brain centers from all somatosensory input. Anesthesia was not continued during the experiment, so that motor control could be studied. Decerebration was considered to be a substitute for anesthesia. Decerebration is now obsolete.

& Udo 1994). Nitric oxide deprivation prevents the cats from learning to walk on the split belt (Yanighara & Kondo 1996), suggesting that cerebellar LTD may be a mechanism for learning this locomotor task.

## THE AGILE MOUSE

Rats and mice may be too small for split-belt treadmills, but they can learn to navigate obstacle courses, which also require good locomotor skills. Skills like ladder climbing and chasm jumping benefit from practice. Acquiring proficiency at these tasks causes increased synaptic density on the spines of P-cells from the paravermal region of lobules IV and V (Kim *et al.* 2002), the same region that receives somatosensory and proprioceptive input from the feet. This indicates that cerebellar synaptic plasticity does occur during learning on an obstacle course.

The "rotorod" task, which requires a small animal to walk along a rotating rod, is another demanding locomotor task that requires practice to perform well. mGluR1 knockout mice are unable to learn the rotorod (Aiba *et al.* 1994), leading many to assume that cerebellar LTD is required for the rotorod task. But in fact, mutants that have lost cerebellar LTD due to deficiencies in glial function can learn the rotorod as quickly as normals (Shibuki *et al.* 1996). The "Erasmus ladder", recently developed in Rotterdam, is another locomotion task that does not require cerebellar LTD. On the ladder, alternate rungs are offset vertically, and the mice learn to walk only on the higher rungs, allowing a measurement of missteps which declines over a few days. Several mutants that fail to express cerebellar LTD have been tested using the Erasmus ladder and none have showed a significant deficit (Schonewille *et al.* 2011). This does not prove that LTD does not normally participate in locomotor learning, but it suggests that other sites and mechanisms of cerebellar plasticity may contribute to learning locomotor tasks, as they do to other kinds of motor learning.

Complex spikes in lobule V vermis, which have somatosensory receptive fields, respond to perturbations[4] of gait, perhaps signaling motor errors (Andersson & Armstrong 1987). However, learning to compensate for perturbations can lead to increased complex-spike as well as simple-spike activity (Lou & Bloedel 1988). In fact, an increase in complex spikes also happens during some reaching tasks, and during pursuit learning (as already mentioned).

[4]The perturbation was caused by a loose rung on the ladder that a cat was walking on. The rung would give way slightly when the cat placed weight on it.

The use of a variety of learning mechanisms by the cerebellum seems to be a common theme across motor systems. There are two points to be made about this variety. First, it suggests a reason for the enormous flexibility of motor systems: the cerebellum has multiple ways of learning things, and the network can (and will) use the most effective mechanism for each situation. Second, the variety of mechanisms means that the cerebellum can tackle fundamentally new problems—including nonmotor ones—with a higher probability that something will work.

## CONCLUSIONS

The posterior lobe of the cerebellar cortex appears to be important for memory encoding for several kinds of skill learning involving the limbs. The exact role of the posterior lobe is less clear, but it may be more crucial when the skill to be learned requires cognition, or explicit learning. On the other hand, the anterior lobe may be required for improving motor performance and for carrying out the task in a coordinated fashion. The cellular mechanisms for skill learning have been little studied, but it seems likely that some kind of cerebellar synaptic plasticity does participate in skill learning, and cerebellar LTD may play some part. Cerebellar LTD does not seem to be required for learning of new locomotor rhythms but if LTD is available, it may participate in such learning.

## REFERENCES

Aiba, A., Kano, M., Chen, C., Stanton, M.E., Fox, G.D., Herrup, K., Zwingman, T.A. & Tonegawa, S. (1994) Deficient cerebellar long-term depression and impaired motor learning in mGluR1 mutant mice. *Cell*, **79**, 377–388.

Andersson, G. & Armstrong, D.M. (1987) Complex spikes in Purkinje cells in the lateral vermis (b zone) of the cat cerebellum during locomotion. *J. Physiol.*, **385**, 107–134.

Aumann, T.D. & Horne, M.K. (1999) Ultrastructural change at rat cerebellothalamic synapses associated with volitional motor adaptation. *J. Comp. Neurol.*, **409**, 71–84.

Baizer, J.S., Kralj-Hans, I. & Glickstein, M. (1999) Cerebellar lesions and prism adaptation in macaque monkeys. *J. Neurophysiol.*, **81**, 1960–1965.

Bastian, A.J. (2006) Learning to predict the future: the cerebellum adapts feedforward movement control. *Curr. Opinion Neurobiol.*, **16**, 645–649.

Bracha, V., Kolb, F.P., Irwin, K.B. & Bloedel, J.R. (1999) Inactivation of interposed nuclei in the cat: classically conditioned withdrawal reflexes, voluntary limb movements and the action primitive hypothesis. *Exp. Brain Res.*, **126**, 77–92.

Diedrichsen, J., Hashambhoy, Y., Rane, T. & Shadmehr, R. (2005) Neural correlates of reach errors. *J. Neurosci.*, **25**, 9919–9931.

Donchin, O., Rabe, K., Diedrichsen, J., Lally, N., Schoch, B., Gizewski, E.R. & Timmann, D. (2012) Cerebellar regions involved in adaptation to force field and visuomotor perturbation. *J. Neurophysiol.*, **107**, 134–147.

Doyon, J., Song, A.W., Karni, A., Lalonde, F., Adams, M.M. & Ungerleider, L.G. (2002) Experience-dependent changes in cerebellar contributions to motor sequence learning. *Proc. Natl. Acad. Sci. USA*, **99**, 1017–1022.

Doyon, J., Penhune, V. & Ungerleider, L.G. (2003) Distinct contribution of the cortico-striatal and cortico-cerebellar systems to motor skill learning. *Neuropsychologia*, **41**, 252–262.

Gilbert, P.F.C. & Thach, W.T. (1977) Purkinje cell activity during motor learning. *Brain Res.*, **128**, 309–328.

Ilg, W., Giese, M.A., Gizewski, E.R., Schoch, B. & Timmann, D. (2008) The influence of focal cerebellar lesions on the control and adaptation of gait. *Brain*, **131**, 2913–2927.

Imamizu, H., Miyauchi, S., Tamada, T., Sasaki, Y., Takino, R., Putz, B., Yoshioka, T. & Kawato, M. (2000) Human cerebellar activity reflecting an acquired internal model of a new tool. *Nature*, **403**, 192–195.

Ito, M. (1993) Movement and thought: identical control mechanisms by the cerebellum. *Trends Neurosci.*, **16**, 448–450.

Jenkins, I.H., Brooks, D.J., Nixon, P.D., Fackowiak, R.S.J. & Passingham, R.E. (1994) Motor sequence learning: a study with positron emission tomography. *J. Neurosci.*, **14**, 3775–3790.

Kim, H.T., Kim, I.K., Lee, K.J., Lee, J.R., Park, S.K., Chun, Y.-H., Kim, H. & Rhyu, I.J. (2002) Specific plasticity of parallel fiber/Purkinje cell spine synapses by motor skill learning. *Neuroreport.*, **13**, 1607–1610.

Kitazawa, S., Kimura, T. & Yin, P.B. (1998) Cerebellar complex spikes encode both destinations and errors in arm movements. *Nature*, **392**, 494–497.

Krakauer, J.W., Ghilardi, M.F. & Ghez, C. (1999) Independent learning of internal models for kinematic and dynamic control of reaching. *Nat. Neurosci.*, **2**, 1026–1031.

Lou, J.S. & Bloedel, J.R. (1988) A study of cerebellar cortical involvement in motor learning using a new avoidance conditioning paradigm involving limb movement. *Brain Res.*, **445**, 171–174.

Lu, X., Hikosaka, O. & Miyachi, S. (1998) Role of monkey cerebellar nuclei in skill for sequential movement. *J. Neurophysiol.*, **79**, 2245–2254.

MacKay, W.A. (1988) Unit activity in the cerebellar nuclei related to arm reaching movements. *Brain Res.*, **442**, 240–254.

Manni, E. & Petrosini, L. (2004) A century of cerebellar somatotopy: a debated representation. *Nat. Rev. Neurosci.*, **5**, 241–249.

Martin, T., Keating, J., Goodkin, H., Bastian, A. & Thach, W. (1996) Throwing while looking through prisms. I. Focal olivocerebellar lesions impair adaptation. *Brain*, **119**, 1183–1198.

Maschke, M., Gomez, C.M., Ebner, T.J. & Konczak, J. (2004) Hereditary cerebellar ataxia progressively impairs force adaptation during goal-directed arm movements. *J. Neurophysiol.*, **91**, 230–238.

Molinari, M., Leggio, M.G., Solida, A., Ciorra, R., Misciagna, S., Silveri, M.C. & Petrosini, L. (1997) Cerebellum and procedural learning: evidence from focal cerebellar lesions. *Brain*, **120**, 1753–1762.

Morton, S.M. & Bastian, A. (2006) Cerebellar contributions to locomotor adaptation during splitbelt treadmill walking. *J. Neurosci.*, **26**, 9107–9116.

Olsen, E. (2012) Blind adventurer trains for his scariest challenge yet. *N.Y. Times*, New York.

Pasalar, S., Roitman, A.V., Durfee, W.K. & Ebner, T.J. (2006) Force field effects on cerebellar Purkinje cell discharge with implications for internal models. *Nat. Neurosci.*, **9**, 1404–1411.

Penhune, V.B. & Doyon, J. (2002) Dynamic cortical and subcortical networks in learning and delayed recall of timed motor sequences. *J. Neurosci.*, **22**, 1397–1406.

Schlerf, J., Ivry, R.B. & Diedrichsen, J. (2012) Encoding of sensory prediction errors in the human cerebellum. *J. Neurophysiol.*, **98**, 54–62.

Schonewille, M., Gao, Z., Boele, H.J., Veloz, M.F.V., Amerika, W.E., Simek, A.A.M., De Jeu, M.T.G., Steinberg, J.P., Takamiya, K., Hoebeek, F.E., Linden, D.J., Huganir, R.L. & De Zeeuw, C.I. (2011) Reevaluating the role of LTD in cerebellar motor learning. *Neuron.*, **70**, 43–50.

Shadmehr, R. & Brashers-Krug, T. (1997) Functional stages in the formation of human long-term motor memory. *J. Neurosci.*, **17**, 409–419.

Shadmehr, R. & Holcomb, H.H. (1997) Neural correlates of motor memory consolidation. *Science*, **277**, 821–825.

Shadmehr, R., Smith, M.A. & Krakauer, J.W. (2010) Error correction, sensory prediction, and adaptation in motor control. *Ann. Rev. Neurosci.*, **33**, 89–108.

Shibuki, K., Gomi, H., Chen, L., Bao, S., Kim, J.J., Wakatsuki, H., Fujisaki, T., Fujimoto, K., Katoh, A., Ikeda, K., Chen, C., Thompson, R.F. & Itohara, S. (1996) Deficient cerebellar long-term depression, impaired eyeblink conditioning, and normal motor coordination in GFAP mutant mice. *Neuron*, **16**, 587–599.

Tseng, Y., Diedrichsen, J., Krakauer, J.W., Shadmehr, R. & Bastian, A. (2007) Sensory prediction errors drive cerebellum-dependent adaptation of reaching. *J. Neurophysiol.*, **98**, 54–62.

Wainscott, S.K., Donchin, O. & Shadmehr, R. (2005) Internal models and contextual cues: encoding serial order and direction of movement. *J. Neurophysiol.*, **93**, 786–800.

Werner, S., Bock, O., Gizewski, E.R., Schoch, B. & Timmann, D. (2010) Visuomotor adaptive improvements and aftereffects are impaired differentially following cerebellar lesions in SCA and PICA territory. *Exp. Brain Res.*, **201**, 429–439.

Wolpert, D.M., Miall, R.C. & Kawato, M. (1998) Internal models in the cerebellum. *Trends Cogn. Sci.*, **2**, 338–347.

Yanagihara, D. & Udo, M. (1994) Climbing fiber responses in cerebellar vermal Purkinje cells during perturbed locomotion in decerebrate cats. *Neurosci. Res.*, **19**, 245–248.

Yanighara, D. & Kondo, I. (1996) Nitric oxide plays a key role in adaptive control of locomotion in cat. *Proc. Natl. Acad. Sci. USA*, **93**, 13292–13297.

## FURTHER READING

Doyon, J., Penhune, V. & Ungerleider, L.G. (2003) Distinct contribution of the cortico-striatal and cortico-cerebellar systems to motor skill learning. *Neuropsychologia*, **41**, 252–262.

Thach, W.T. & Bastian, A.J. (2004) Role of the cerebellum in the control and adaptation of gait in health and disease. *Prog. in Brain Res.*, **143**, 353–366.

Shadmehr, R., Smith, M.A. & Krakauer, J.W. (2010) Error correction, sensory prediction, and adaptation in motor control. *Ann. Rev. Neurosci.*, **33**, 89–108.

Shmuelof, L. & Krakauer, J.W. (2011) Are we ready for a natural history of motor learning? *Neuron*, **72**, 469–477.

# Section III

## Precision Control

We need precisely timed, coordinated movements, not only for high-end sports competitions, but also for everyday life. But precision requires processing power. In situations where large amounts of information need to be processed quickly the cerebellum assists the cerebral cortex by performing fast computations, much as using Google instead of the library can speed up a search.

Motor memories are kept in the cerebellum for a long time, and it expresses them, as we have seen. But the cerebellum also adjusts movements in real time. Some of the same error-correction circuitry that is used in learning also participates in making the real-time adjustments. We can visualize the cerebellum as the general manager of the motor system, taking concepts and converting them into action plans. Both for error correction and for movement planning, the cerebellum must interface and cooperate with the cerebral cortex.

The cerebellum is a sensorimotor structure. Purkinje cells (P-cells) respond not only to intended movements but also to sensory inputs, including the sensory signals that evoke the movement and the sensory feedback that allows the movement to be perfected. Given enough time, the cerebellum can learn adjustments for mechanical loads and for muscle weakness, to correct inaccurate gaze-stabilizing movements, saccades, cursor movements and throws, and the timing of a blink. If it can do all of that, then perhaps it can correct similar errors immediately, or with only a small delay. This would provide temporary solutions for unfamiliar or challenging situations.

Patients with cerebellar damage are often uncoordinated; they have difficulty walking, and might not pass a standard sobriety test. Their many deficits force us to recognize just how many motor tasks require cerebellar memory and/or computational capacity. Over the years, there have been many studies of the role of the cerebellum in precision movements. In Chapters 8–10, we will take a short walk around this thoroughly mapped territory.

---

*The Cerebellum: Learning Movement, Language, and Social Skills*, First Edition. Dianne M. Broussard.

# 8 Coordination

Coordinated movements, especially those that have been practiced extensively, can seem almost miraculous. Think of the blind kayaker, Olympic synchronized divers, baseball players, and the millions of examples in nature. Watch your own hand as you pick up a pen. The movement is not understood; it just happens. It is a mystery.

Movements of the hands are especially precise in humans, and this has often been given as the reason for the large cerebellar hemispheres and dentate nucleus in our species. Because humans and other primates primarily use vision for gathering information about the world, eye movements are also especially important in our species. In both cases, the cerebellum is used to achieve a precise spatial relationship between the hand or the eye and the outside world. In doing this, the cerebellum needs to use an external frame of reference. This means that movements are defined with respect to objects outside the body, and requires planning of the movement within the external frame, in what we usually call kinematic or sensory coordinates.

Sensory coordinates depend on the sensory system. For example, in the visual system, the frame of reference is the retina, and it moves with the eye. Visual information must be converted to an external frame of reference in order to be combined with other sensory information. But in a movement plan, each muscle has its own clearly defined contribution. Coordination requires the movement to be represented very differently, in motor coordinates, before it can be carried out. The transformation from one coordinate system to another is computationally intensive and may make use of the cerebellum.

A second problem has to do with multiple joints. "Decomposition of movement," in which the patient substitutes sequential rotations for simultaneous rotations around multiple joints, is a well-known sign of cerebellar damage. Why? By a simple analogy, simultaneous rotations present a complicated mathematical problem, and are best avoided if the computer is down. An accurate reaching movement can be accomplished in a large—actually, infinite—number of ways. We must somehow choose

*The Cerebellum: Learning Movement, Language, and Social Skills*, First Edition. Dianne M. Broussard.

among an infinite number of possible combinations of muscular contractions. Timing the contractions of different muscles that will act together, particularly the muscles that are acting at different joints, is both difficult and absolutely necessary. This brings us to the third problem to be solved by coordination, getting the muscles to work together. In order to contribute to this task, the cerebellum needs access to the muscles. P-cells could control particular muscles, and correctly time their activation or relaxation to execute the movement plan. Of course, the motor cortex also has direct access to the muscles. The forebrain can execute movements, and it can do it without the cerebellum. But here is the key point: the forebrain does a poor job of it. In reaching for the branch, we would miss it.

Finally, a movement plan must have a learned component and a component that is flexible. The flexible component is the response to feedback that occurs during the movement. The learned component is really an algorithm, stored as synaptic weights and neuronal excitability, which can be used by the cerebellum to generate the desired movement based on the input from the forebrain. These algorithms could serve to assemble components of some kind into full-fledged movement plans. A small number of movement components could, theoretically, produce a very large number[1] of combinations and a large repertoire of actual movements. In some cases they might correspond to the so-called "movement primitives," or very simple movements (Flash & Hochner 2005). Or, the components might be rotations around individual joints. Or, they could even be individual muscle contractions. The nature of the components varies depending on the situation.

As an example of how a movement plan is put together, let us examine eye movements.

## PRECISE OCULAR COORDINATION

Before reaching for the tree limb that will break our fall, we need to see it and judge its distance: this is the basis for "eye–hand coordination." Eye–hand coordination really requires two coordinated movements (first the eyes, then the hand) plus a distance measurement, so that we know how far to reach. The best cue for the distance of an object from the eyes is retinal disparity. The image of the tree limb falls on slightly different parts of the two retinas, and the difference (i.e., disparity) is used to calculate

[1]The number of combinations is $2^n$, where $n$ is the number of components. This means that only 32 components that are compatible with each other could yield over 4 *billion* combinations. Of course, not all components would be compatible in this way.

the distance. To know the distance we need two eyes, pointing at the same place, so that we have good depth vision. All eye movements must be precisely coordinated between the two eyes so that they both stay on the same target. This requires the cerebellum. Double vision due to ocular misalignment is a characteristic of cerebellar damage and degeneration (Versino *et al.* 1996).

Physiologists have identified a few "common signals" that go from the brainstem to both eyes so that they can move in synchrony. Copies of some motor signals even cross the midline of the brainstem to reach the correct muscle that will move the other eye in the same direction. But for the most part the two eyes are controlled independently (King 2011), and we must learn to coordinate them for different tasks. We need a lot of flexibility in this, because exactly what needs to be done depends on the path taken by the visual object. A tracking movement for an object moving across the visual field is similar in both eyes; this is called a "conjugate" movement. In contrast, an object moving directly toward or away from us requires opposite tracking movements in the two eyes. The coordinated movement of both eyes in opposite directions is called vergence. There are two kinds of vergence movements: convergence, which is required if the object is getting closer, and divergence, which is what we need if the object is moving away.

The cerebellum is necessary for normal performance of both vergence movements and conjugate eye movements (Sander *et al.* 2009). The dorsal vermis (lobules VI and VII) and the cerebellar hemispheric regions lateral to it (some areas in crus I and crus II) are activated for both (Richter *et al.* 2004), and the flocculus and ventral paraflocculus also participate. But there may also be separate areas of the cerebellar cortex that are responsible for programming vergence and conjugate eye movements (Sander *et al.* 2009). Ischemic lesions of the cerebellum often (but not always) interfere selectively with divergence movements, when the eyes are moving away from each other to track a target that is moving away (Sander *et al.* 2009). On the other hand, injection of muscimol into the vermis impairs convergent tracking selectively (Nitta *et al.* 2008). In the deep cerebellar nuclei (DCN), the fastigial and interpositus nuclei act together to regulate the vergence angle. Some neurons in the nucleus interpositus respond only during divergence, and microstimulation of this region can cause the eyes to diverge (Zhang & Gamlin 1998). All of this indicates that the cerebellum plays a part in carrying out the vergence components of movements.

Both for vergence and for conjugate movements, the cerebellum keeps both eyes on the target by functioning as a servocontroller. A servocontroller is a wonderfully helpful device that compares the output of a system with what it is supposed to produce. The difference, called an error signal, is then used to generate a corrective change in the output of the system. Cerebellar servocontrollers can stabilize gaze with respect to the world as well as follow interesting objects. For stabilizing gaze, the error signal needs to come from the entire visual field, while for tracking ("smooth pursuit"), it should come only from the object to be tracked.

The flocculus and ventral paraflocculus stabilize gaze using the optokinetic response, which works like a servocontroller using direct visual feedback from the eyes. Smooth pursuit eye movement, including both its vergence and conjugate components, is also guided by a servocontroller. The error signal and the motor signals for pursuit are both represented in the discharge patterns of P-cells, both in the ventral paraflocculus (Lisberger & Fuchs 1978; Stone & Lisberger 1990) and in lobules VI and VII of the vermis (Suzuki *et al.* 1981; Suzuki & Keller 1988a). Whenever there is motion of the image of the chosen target on or near the fovea, a signal arrives in the cerebellar cortex on mossy fiber pathways. This signal arises from the visual cortex. Other kinds of signals, such as the speed of a target that is imagined rather than seen, can also drive pursuit.

Another key part of the servocontroller is motor feedback (i.e., positive feedback), which may come from an internal model of the eyeball and associated tissues. This signal is also a part of any servocontroller. Put simply, the error signal alone is not enough to generate a motor command, because there is not always an error. The positive feedback signal supports ongoing movements.

Given the output of the servocontroller, ccordination requires orchestrating the movement at the level of motoneurons. During pursuit, P-cells in the flocculus and ventral paraflocculus inhibit premotor neurons in the vestibular nuclei that inhibit the motoneurons. Because they have different on-directions, different groups of P-cells are responsible for accelerating and decelerating each movement. When the P-cells fire during pursuit in their on-direction, muscle tension will increase. At the same time, P-cells whose off-direction is the direction of pursuit must reduce their firing rates, in order to reduce muscle tone in the antagonist muscle. When the pursuit target stops abruptly, the antagonist group of P-cells shows a transient increase in firing (Krauzlis & Lisberger 1994), suggesting that they actively brake the movement via tension in the antagonist muscles.

During pursuit, the cerebellar signal represents eye movement in motor coordinates. Both sensory and the motor signals for pursuit are encoded by P-cells in the ventral paraflocculus (Lisberger & Fuchs 1978; Stone & Lisberger 1990) and posterior vermis (Suzuki *et al.* 1981; Suzuki & Keller 1988a). However, many of them respond during pursuit that is directed downward and slightly toward the midline of the body, which coincides with the pulling direction of the inferior rectus muscle. In fact, microstimulation in the ventral paraflocculus causes mostly slow, pursuit-like, downward eye movements. Other P-cells increase firing during pursuit straight toward the side that the P-cells are on, which coincides with another muscle, the lateral rectus.

Although pursuit and saccades are mostly guided by different aspects of the visual signal (position error for saccades, and velocity error for pursuit), similar error signals can also contribute to both, and there is a great deal of overlap in their cerebellar representation. Lesions of the cerebellar cortex, particularly the ventral paraflocculus, reduce the gain of pursuit (Rambold *et al.* 2002), and lesions of the dorsal vermis result in dysmetric, or inaccurate, saccades. However, lobules VI and VII of the vermis contribute to both saccadic and pursuit eye movements (Krauzlis & Miles 1998). Neurons in the fastigial nucleus also respond to both types of movement.

## Stopping on target

Obviously, a really good movement plan should contain some instructions about stopping. In fact, stopping any rapid orienting movement precisely on target requires the cerebellum. Dysmetria is often seen in cerebellar patients that are trying to move the arm rapidly to point at a target. When the hand moves across the body in a pointing task, for example, from the right to the left, it tends to go too far before stopping ("past pointing"). The same is true of saccades. In many cases, saccades overshoot following cerebellar damage; undershooting saccades are also seen (Versino *et al.* 1996). Saccades have been very thoroughly studied, and they can illustrate some possible ways of generating an accurate orienting movement.

How are saccades stopped? First of all, sensory feedback can be ruled out. Servocontrollers, at least those relying on vision, are too slow to stop a saccade on time. Instead, feedback from an efference copy of the motor signal, supplemented by proprioceptive feedback, is used to determine when the saccade should be stopped. Next, the cerebellum puts on the brakes. P-cells on the side that the saccade is directed toward generate a burst of spikes before and during the saccade. This discharge causes a pause in fastigial neurons during the saccade (Ohtsuka & Noda 1995). After the P-cell burst ends, the fastigial neurons generate a burst of firing—due to rebound excitation—that terminates the saccade. Without the fastigial burst, the saccade still ends, but it does not end on target. Burst firing by the fastigial neurons triggers a braking mechanism, possibly involving inhibition of agonists combined with excitation of antagonist muscles. In other words, the cerebellum stops the movement by coordinated actions of the muscles involved.

After saccade size has been adjusted by motor learning, whether a particular saccade amplitude is used or not depends on whether the current circumstances match the context in which it was learned (Shelhamer & Clendaniel 2002). This selective implementation requires planning. Adaptation to prisms is also applied depending on the context. Similar strategies may be used for other kinds of movements.

In summary, lobules VI and VII of the vermis may coordinate agonist activation with antagonist inhibition during saccades. It may also coordinate agonist inhibition with antagonist activation to stop the saccade. Similar strategies may be used for other orienting movements, such as pointing.

## COORDINATING THE EYES AND THE HEAD

Head and eye rotations can both contribute to a combined tracking or orienting movement. This strategy extends the range of possible eye positions and speeds, because head and eye motion add. Coordinating the eyes and the head turns tracking into a multi-joint problem (i.e., a movement with multiple centers of rotation). Since we know a lot about eye movements, I will examine eye–head coordination in more detail to illustrate several key points about multi-joint movements.

The representation of eye and head movements in the cerebellum depends on whether the head movement is active or passive. In retrospect this seems obvious, since sensory feedback that is due to active movements is suppressed in many systems. For the vestibular system, which senses head movements, a passive head movement[2] generates a much smaller response (in some vestibular neurons) than an active one does (McCrea *et al.* 1999). This difference is useful, because if we are making an active movement, we do not want to generate stabilizing reflexes of the eyes and head that would undo our action. The distinction between active and passive movements could be made by a servo-like mechanism, an internal comparator. What is being compared in this case is a copy of the command signal to the neck muscles—known as efference copy—and the feedback sensory signal from stretch receptors in the neck muscles (Roy & Cullen 2004). The comparison may occur in the cerebellum. Activity in the fastigial nucleus seems to represent the passive component of the motion, suggesting that the comparison is done in the cerebellar vermis (Angelaki & Cullen 2008).

If the VOR is fully functional, eye and head movements cannot add. This means that if we want to coordinate the eyes and the head we need to suppress the VOR, at least until the eyes are on target. In fact, the VOR does not operate during either saccades or tracking. The exception is that once the eyes are on target, if the head continues to move because of its inertia, then the VOR does function. The cerebellar cortex has been proposed to have a role in VOR cancellation. During passive head movements, P-cells in the ventral paraflocculus, flocculus, and posterior

[2]The passive head movements in these experiments are imposed by a torque motor.

vermis encode a sum of eye and head motion, called gaze velocity (Lisberger & Fuchs 1978; Suzuki & Keller 1988b). Using gaze velocity, the eye movement can be coordinated with, or take into account, the passive head movement. The flocculus and ventral paraflocculus are involved in this kind of coordination (Rambold *et al.* 2002). The gaze velocity signal in floccular P-cells, which is inhibitory on VOR interneurons, may suppress the VOR allowing the eyes and head to move together.

Does the cerebellum exert low-level control over somatic motoneurons, as it does for extraocular ones? At least for the neck muscles involved in head movements, this appears likely. For saccades that require movements of the eyes and head, the vermis and fastigial nucleus are involved in both. Inactivation of the fastigial nucleus results in hypermetria for gaze shifts toward the side the nucleus is on, and hypometria for gaze shifts in the opposite direction. When the head moves freely and contributes to the saccade (or gaze shift), the caudal fastigial nucleus responds, just as it would for saccades (Fuchs *et al.* 2010). Thus, it appears that the fastigial nucleus may control both eye and head components of the gaze shift using similar signals.

What about planning the gaze shift? We use the neck muscles or do not use them, depending on the situation; for example, we are more likely to use neck muscles in a situation where we are going to need to run in the new direction, or grasp something, or make another gaze shift in the same direction (Monteon *et al.* 2012). We need to decide this in advance, because accelerating the head takes more time than accelerating the eyes. It would be interesting to investigate whether the cerebellum plays any part in the decision or not.

## ERROR CORRECTION FOR LIMB MOVEMENTS

The cerebellum also exerts low-level control over limb motoneurons. It can access them both by way of the interpositus/red nucleus pathway and by way of the fastigial/lateral vestibular nucleus pathway, as well as through the motor cortex. Stimulation of the cerebellar cortex causes movement, revealing a motor map that resembles a collection of body parts more than a coherent representation of the body. Nevertheless, the map is functional. Stimulation of the hemispheric portion of lobules IV and V evokes forelimb movements. In the posterior lobe, stimulating lobule VI evokes movements of the face and jaw, and at least some parts of crus II are associated with forelimb movements. The finger area is not found in the same lobule as the arm, but slightly further caudal, in lobule VIII (Manni & Petrosini 2004). The regions of the inferior olive that are associated with the anterior lobe are important for coordinated locomotion and accurate reaching by particular muscles (Horn *et al.* 2010).

Sensors in the joints and tendons detect actual movement and provide information to the cerebellum. The sensory inputs to P-cells are somatotopically organized. The anterior lobe of the cerebellum contains a proprioceptive and a somatosensory representation of the entire body, with the forelimbs in lobules IV and V. The posterior lobe also contains a (very distorted) somatotopic arrangement. The illusion of movement can be generated by vibrating the tendons, and this illusion also causes cerebellar activity in restricted regions of the posterior lobe. This arrangement facilitates comparisons of intended and actual movement, both for servocontrol and for distinguishing active from passive movement.

Servocontrollers are as important for limb movements as they are for eye movements. Servocontrol can enable the limbs to respond to changes in external forces due to the inertia of an object that we are trying to move around, or of the limb itself. Cerebellar patients may not be able to hold the positions of joints against gravity, and this is something servocontrol can help them to do.

Cerebellar lesions also interfere with compensation for changes in load. Nearly a century ago, Gordon Holmes, a British neurologist, studied a large group of soldiers with gunshot wounds to the cerebellum. Among many important observations, Holmes described a characteristic feature of cerebellar damage: when a weight that the patient had been holding by a string was suddenly detached by cutting the string, the arm would fly upwards. The patient would be unable to stop the arm and sometimes ended up hitting himself in the face (Holmes 1922).

The role of the cerebellum in servocontrol may also explain intention tremor. Cerebellar patients and animals with cerebellar lesions have a pronounced tremor that appears during movements (as opposed to at rest). Cerebellar tremor is consistent with a fast cerebellar control loop that, when damaged, exposes another, slower loop in parallel with it. The slower loop might require the stretch reflex, for example. During the time required for activity to travel around the loop, the activity level can change significantly because the loop is slow. These changes reverberate around the loop, resulting in an oscillating signal and tremor. If this hypothesis is correct, then the cerebellum may normally prevent tremor by speeding up our response to small errors.

The interpositus nucleus has a role in feedback control of the limbs. In a task where monkeys are required to correct for external perturbations of hand position, the interpositus responds before the correction happens (Thach *et al.* 1992). The interpositus might actually generate the correction, by means of its projections to the red nucleus and to the region of the thalamus that provides input to motor cortex. It could be making corrective changes to the motor command.

The cerebellar servocontroller has been proposed to coordinate movements as part of a so-called "equilibrium point" control system, which would allow computation of corrective signals at discrete times. But attempts to model movement using equilibrium points have failed, probably because they oversimplify the mechanics of the system (Gomi & Kawato 1996). Also, depending on how errors are fed back, servocontrol may be intrinsically too slow to coordinate muscles during a complex movement. We need another way of implementing the movement plan.

## PLANNING FOR MULTIPLE JOINTS

For reaching and throwing movements, the goal of the movement and the motor command must be represented using different coordinate systems: spatial coordinates, which define the sensory world, and motor coordinates, which define the body's own reference frame. In order to plan a movement, signals must be converted from the spatial reference frame to motor coordinates, so that individual muscles can be given their tasks (when and how much to contract). Some movements can be carried out using many possible combinations of muscles; there are many ways to throw a dart at a target, for example. This means that the conversion from sensory to motor coordinates can be done in many possible ways.

Some regions of the cerebellum seem to operate in motor coordinates (i.e., the anterior lobe and the regions that project to the interpositus, and some parts of the posterior lobe), while others may operate in a spatial reference frame for motor planning purposes (other parts of the posterior lobe hemisphere and dentate nucleus). This is consistent with using the cerebellum to do the conversion from sensory to motor coordinates.

Throughout the cerebellar cortex, microzones are narrow parasagittal zones that receive input from climbing fibers with similar sensory receptive fields[3] in the periphery and project back, via the DCN, to the same regions of the inferior olive. This means that some of the microzones in the posterior lobe are associated, through their climbing fibers, with particular regions in the outside world. The spatial, or external, reference frame is also represented in the parietal and prefrontal cerebral cortex, and one region of the posterior lobe of the cerebellum—crus II—is interconnected, through the ventral dentate nucleus, with parietal and prefrontal cortex (Glickstein *et al.* 2011). Crus II is lateral to lobules VI and VII, and is relatively large in humans. These connections are consistent with the idea that crus II may function in spatial perception and/or motor planning with respect to the outside world.

Crus I, another region in the hemispheres of the posterior lobe, may have its own higher-level function in motor planning. A small region of crus I responds during the illusion of wrist movement that is caused by

[3]A neuron's sensory receptive field is the region in space, or on the body, where the neuron can detect a change in the environment.

tendon vibration, and so do supplementary motor areas and the cingulate motor area of the cerebral cortex. The activation in crus I can also be modified by an imagined wrist movement, along with activation of the dorsal premotor cortex (Naito *et al.* 2002). Crus I may be involved in a comparison of imagined, or predicted, and sensed movement. It also seems to be necessary for converting error feedback signals from one frame of reference to another, for motor planning, and for immediate correction of movements (Taig *et al.* 2012). The motor frame of reference is based on the axes of rotation for each joint, in the same way that Cartesian coordinates are based on x, y, and z axes. Because of the number of joints and muscles involved, the representation of motor signals must be extremely complex.

The dentate nucleus operates both in sensory and in motor coordinates. As already mentioned, the ventral dentate nucleus is interconnected with the prefrontal and parietal regions of the cerebral cortex (Dum & Strick 2003). The ventral dentate receives abundant input from the association areas of the cortex by way of the pons. The dentate nucleus generates an early signal before a visually guided hand movement, which means that it could be involved in organizing the contributions of the muscles ahead of time (Thach *et al.* 1992).

## *Executing the movement plan*

The arm regions of the motor cortex receive motor signals, through the dorsal dentate nucleus, from lateral lobules IV and V in the cerebellar anterior lobe (Glickstein *et al.* 2011). These projections are likely involved in the lower-level details of the movement plan. The fastigial nucleus, nucleus interpositus, and dorsal dentate nucleus each has its own representation of the body. Activity in the dentate and interpositus nuclei may converge on the same downstream targets in the thalamus, which connect with motor cortex (Thach *et al.* 1992). Subsets of dentate and interpositus neurons both are active during contraction of each muscle group (Wetts *et al.* 1985). The dentate and interpositus seem to share the task of coordinating muscles that act on a given joint (Harvey *et al.* 1979).

One possibility is that the cerebellum quickly plans the details of a movement that has been "roughed out" by the cortex. In this scenario, patients that have lost the relevant part of the cerebellum would be forced to consciously plan every movement detail using the forebrain. This might take longer, and in fact, cerebellar patients do take longer to move. Holmes' cerebellar gunshot patients avoided using their affected limbs if they could (using the other limbs instead). When forced to use the affected limbs they tended to move slowly (Holmes 1917), as do cerebellar stroke patients (Timmann *et al.* 1999; Taig *et al.* 2012). Cerebellar stroke patients are slower to accelerate rotations at the elbow, wrist, and finger joints, and also slower to stop; in general, arm movements tend to overshoot (Hore *et al.* 1991).

To plan a coordinated movement with simultaneous rotation around multiple joints, the cerebellum does a multitude of calculations. Without the cerebellum it is better not to try it. In fact, when they are not rushed, patients with cerebellar damage tend to break down a reaching movement into single-joint movements (Bastian *et al.* 1996), giving themselves time to solve the problem using slower cortical circuits.

Throwing is a good example of a coordinated movement that requires a detailed plan. The most important feedback on a throw happens after it is over, and what is learned may actually be a plan for coordinating the muscles during stereotyped throwing patterns. Cerebellar patients cannot throw a ball accurately. They also have more variable trajectories (for the ball, that is) than normal subjects (Timmann *et al.* 1999). This suggests that without good cerebellar function, they tend to use a varying set of muscles as opposed to learning a stereotyped plan.

An important detail of any movement is the optimal amount of co-contraction of the agonist and antagonist muscles. Co-contraction stiffens the joint, but also makes it stronger. Often, Holmes' soldiers could not exert as much force as they thought they should be able to exert (Holmes 1917), suggesting they used too little co-contraction. The optimal amount for a given movement depends on the movement's amplitude and speed. This information must be available ahead of time so that it can be used in planning. The cerebellum also needs access to known (i.e., previously learned) properties of the arm, hand, and independent fingers, and there are many such characteristics. Dealing with this large amount of quantitative information, which has been acquired by trial and error, could be one reason for the expansion of the posterior lobes in humans.

Inactivation of the dentate nucleus in monkeys results in difficulties with coordinated, visually guided, or memorized reaching movements, reminiscent of some of Holmes' gunshot patients. In both cases, movements around both the shoulder and the elbow tend to be excessive, which results in overshooting the target (Thach *et al.* 1992). In addition to problems with co-contraction, cerebellar patients also do not produce appropriate muscle torques during fast reaching. The problem is as follows: To make a coordinated multi-joint movement, the plan for movement at each joint has to predict the torques that the movement is going to cause at the other joints, called interaction torques. The ability to predict interaction torques is impaired in cerebellar lesions, and the loss of this capability is probably an important factor in ataxia of the arm (Bastian *et al.* 1996).

## Using the hands

The dentate nucleus has a "toothed" appearance (hence the name) in the hominid family of great apes and humans. This is because during the evolution of the hominids, the dentate nucleus changed from a thick, folded layer into a thinner layer with a large surface area and numerous folds.

Also, the number of cells increased. In macaques, the dentate nucleus contains roughly 68,000 neurons, but in humans, it contains 3.5 *million* (Glickstein *et al.* 2011). Why? Clearly, the dentate must have some functions that are definitively human. One such function is to coordinate the hands.

Hand and finger movements can be unbelievably fast and accurate. Piano virtuosos can play at what seems an impossible speed with perfect accuracy. Such precise finger coordination and timing is clearly an appropriate job for the cerebellum. In fact, even something simple like opposing the index finger and thumb, for the precision grip, is impossible for humans without the cerebellum. Simple finger movements involve two main regions of the cerebellar cortex, one in the anterior lobe (lobules IV/V) and one in the posterior lobe (lobule VIII). Both are in the lateral cerebellum (Desmond *et al.* 1997). In monkeys, the dentate nucleus is required for "precision pinch" (Thach *et al.* 1992) and for coordination of the fingers and wrist, for grasping (Holmes 1939).

The cerebellar gunshot patients examined by Holmes often had difficulty keeping fingers that were not needed for a particular movement out of the way (Holmes 1939). Fingers tend to move together, as do toes. For one thing, finger tendons have mechanical connections, so that fingers tend to move passively when the next finger over is moved. In addition, some muscles control multiple fingers, so that motor units will actually move them together. Finally, muscles are often coactivated. The result of all this is known as "force enslaving." In a recent study of stroke patients, cerebellar damage significantly increased the amount of enslaving (Brandauer *et al.* 2012). For typing and playing the piano, one task of the cerebellum may be to free the enslaved fingers. To do this, motor commands could include a stabilizing command for fingers that are not involved in a particular movement.

In summary, although coordination of agonist and antagonist for single-joint movements is impaired in cerebellar patients, it is clear that multi-jointed movements and movements that are coordinated across limbs, or fingers, are more severely impaired. This may be because in multi-jointed movements, there are more interaction torques that are unaccounted for and hence more sources of variability. It may also reflect force enslaving, especially for finger movements. The slower movements in cerebellar patients are probably carried out by other motor systems (i.e., the cerebral cortex and basal ganglia) that normally act in parallel with the cerebellum, but that are less efficient at performing calculations.

## INTERNAL MODELS REVISITED

Good planning always involves predicting consequences; poor predictions mean that the movement will need corrections by servocontrollers, or will fail. In order to plan a movement accurately, we make an educated guess at what the motor commands will do. Interaction torques, for example, must be predicted. This is done using both dynamic and kinematic models of the mechanical system that we want to move (the "plant" in engineering terms).

A linear filter is a simple device that may be used by motor control systems in many situations. It generates an output that is always proportional to its input. A forward dynamic model is an algorithm, which can be a linear filter or something similar, containing information about the characteristics of the plant (in our example, the arm). The algorithm calculates what the arm will do when the command reaches it. In a sense, the forward model is a virtual arm, plus whatever loads are attached to it. One task for a forward model would be to generate the right signal to counteract force enslaving in situations where we want to move the fingers independently; another is to predict interaction torques; still another, to compensate for anticipated loads.

The forward model can also be used for error correction: it can estimate the hand position, for example, with much less delay than visual feedback from the actual hand would require, allowing for fast online corrections, and also for stopping the movement accurately. Saccades, for example, may stop on target using reafference signals that are processed by internal models. The cerebellum has also been proposed to contain forward models that act to predict the sensory feedback that should be obtained from the movement. This prediction can then be compared with the actual feedback[4] (Wolpert *et al.* 1998).

Inverse models of motor systems are also used in motor control. Inverse dynamic models do the opposite, in terms of dynamic filtering, as the plant. For example, if the plant has a very high inertia, the inverse model filters the signal so that it starts with a sudden pulse of acceleration, essentially giving the plant a big push. In general, inverse models produce the correct motor command to deal with the object's characteristics.

Cerebellar damage can prevent the eyes from remaining on target after a saccade. The eyes are also unstable in general; they tend to drift upward, and be continually repositioned in a "downbeat nystagmus." Some P-cells maintain a steady discharge, correlated with eye position, after the end of a saccade. The nystagmus could be explained by the loss of an eye-position signal in P-cells of the flocculus and ventral paraflocculus. This tonic signal is the output of an inverse model of the eyeball and extraocular muscles (the oculomotor plant).[5] The function of the tonic signal is to

[4]This is a very simplified description of a Smith predictor. For a more detailed description and a discussion of forward and inverse models see Wolpert *et al.* (1998).

[5]The inverse model includes what is known as the "neural integrator" described in Robinson (1981).

maintain eye position against the elasticity of the muscles. Thus the output of the cerebellum is an appropriate, "plant-compensated" command signal for controlling the speed of the eyes. Compensation for plant dynamics, such as inertia, is built into the signal.

Inverse kinematic models are also useful, for example, in corrections to a throw that are based on visual monitoring of the arm. Among cerebellar patients, those with more damage to crus I are more impaired in making corrections to reaching movements based on looking at the arm. Although the sensory feedback information is visual, in this case, it is in the wrong frame of reference for correction of the motor command. Crus I seems to be responsible for the inverse transformation from motor coordinates to spatial coordinates (Taig *et al.* 2012).

If we keep several different inverse models on hand for different contexts (for example, prisms or no prisms) the motor system can become extremely flexible. Daniel Wolpert has pointed out that if we have 32 inverse models and each makes a contribution, or not, to the motor command, that gives us $2^{32}$ possible combinations of inverse models. These internal models are possible building blocks for behavior (Wolpert *et al.* 1998), just as others have proposed that "motor primitives" are. The exact nature of the building blocks may depend on the situation and may also be different in different cerebellar regions.

## CONCLUSIONS

The cerebellum performs many functions that we need for interacting with our environment. It guides movements precisely and accurately in relation to objects in the outside world, translating motor commands from external reference frames to internal ones that reflect our joints and muscles. It implements the plan for the movement, including stopping it at the right time. It takes into account joint interactions, force enslaving, motor strategies, and other important factors, to plan the movement. It uses internal models of our bodies to compensate for their physical characteristics. Finally, it can correct movements rapidly using both internal models and sensory feedback. Thanks to the cerebellum, we can perform astonishing feats of precision control for everyday living, for survival, and to enrich our experience.

## REFERENCES

Angelaki, D.E. & Cullen, K.E. (2008) Vestibular system: the many facets of a multimodal sense. *Ann. Rev. Neurosci.*, **31**, 125–150.

Bastian, A.J., Martin, T.A., Keating, J.G. & Thach, W.T. (1996) Cerebellar ataxia: abnormal control of interaction torques across multiple joints. *J. Neurophysiol.*, **76**, 492–509.

Brandauer, B., Hermsdorfer, J., Geissendorfer, T., Schoch, B., Gizewski, E.R. & Timmann, D. (2012) Impaired and preserved aspects of independent finger control in patients with cerebellar damage. *J. Neurophysiol.*, **107**, 1080–1093.

Desmond, J.E., Gabrieli, J.D.E., Wagner, A.D., Ginier, B.L. & Glover, G.H. (1997) Lobular patterns of cerebellar activation in verbal working-memory and finger-tapping tasks as revealed by functional MRI. *J. Neurosci.*, **17**, 9675–9685.

Dum, R.P. & Strick, P.L. (2003) An unfolded map of the cerebellar dentate nucleus and its projections to the cerebral cortex. *J. Neurophysiol.*, **89**, 634–649.

Flash, T. & Hochner, B. (2005) Motor primitives in vertebrates and invertebrates. *Curr. Opin. Neurobiol.*, **15**, 660–666.

Fuchs, A.F., Brettler, S. & Ling, L. (2010) Head-free gaze shifts provide further insights into the role of the medial cerebellum in the control of primate saccadic eye movements. *J. Neurophysiol.*, **103**, 2158–2173.

Glickstein, M., Sultan, F. & Voogd, J. (2011) Functional localization in the cerebellum. *Cortex*, **47**, 59–80.

Gomi, H. & Kawato, M. (1996) Equilibrium-point control hypothesis examined by measured arm stiffness during multijoint movement. *Science*, **272**, 117–120.

Harvey, R.J., Porter, R. & Rawson, J.A. (1979) Discharges of intracerebellar nuclear cells in monkeys. *J. Physiol.*, **297**, 559–580.

Holmes, G. (1917) The symptoms of acute cerebellar injuries due to gunshot injuries. *Brain*, **40**, 461–535.

Holmes, G. (1922) On the clinical symptoms of cerebellar disease and their interpretation. *Lancet*, **199**, 1177–1182.

Holmes, G. (1939) The cerebellum of man. *Brain*, **62**, 1–30.

Hore, J., Wild, B. & Diener, H.C. (1991) Cerebellar dysmetria at the elbow, wrist, and fingers. *J. Neurophysiol.*, **65**, 563–571.

Horn, K.M., Pong, M. & Gibson, A.R. (2010) Functional relations of cerebellar modules of the cat. *J. Neurosci.*, **30**, 9411–9423.

King, W.M. (2011) Binocular coordination of eye movements—Hering's Law of equal innervation or uniocular control? *Eur. J. Neurosci.*, **33**, 2139–2146.

Krauzlis, R.J. & Lisberger, S.G. (1994) Temporal properties of visual motion signals for the initiation of smooth pursuit eye movements in monkeys. *J. Neurophysiol.*, **72**, 150–162.

Krauzlis, R.J. & Miles, F.A. (1998) Role of the oculomotor vermis in generating pursuit and saccades: effects of microstimulation. *J. Neurophysiol.*, **80**, 2046–2062.

Lisberger, S.G. & Fuchs, A.F. (1978) Role of primate flocculus during rapid behavioral modification of vestibuloocular reflex. I. Purkinje cell activity during visually guided horizontal smooth-pursuit eye movements and passive head rotation. *J. Neurophysiol.*, **41**, 733–763.

Manni, E. & Petrosini, L. (2004) A century of cerebellar somatotopy: a debated representation. *Nat. Rev. Neurosci.*, **5**, 241–249.

McCrea, R.A., Gdowski, G.T., Boyle, R. & Belton, T. (1999) Firing behavior of vestibular neurons during active and passive head movements: vestibulo-spinal and other non-eye-movement related neurons. *J. Neurophysiol.*, **82**, 416–428.

Monteon, J.A., Avillac, M., Yan, X., Wang, H. & Crawford, J.D. (2012) Neural mechanisms for predictive head movement strategies during sequential gaze shifts. *J. Neurophysiol.*, **108**, 2689–2707.

Naito, E., Kochiyama, T., Kitada, R., Nakamura, S., Matsumura, M., Yonekura, Y. & Sadato, M. (2002) Internally stimulated movement sensations during motor imagery activate cortical motor areas and the cerebellum. *J. Neurosci.*, **22**, 3683–3691.

Nitta, T., Akao, T., Kurkin, S. & Fukushima, K. (2008) Involvement of the cerebellar dorsal vermis in vergence eye movements in monkeys. *Cereb. Cortex*, **18**, 1042–1057.

Ohtsuka, K. & Noda, H. (1995) Discharge properties of Purkinje cells in the oculomotor vermis during visually guided saccades in the macaque monkey. *J. Neurophysiol.*, **74**, 1828–1840.

Rambold, H., Churchland, A., Selig, Y., Jasmin, L. & Lisberger, S.G. (2002) Partial ablations of the flocculus and ventral paraflocculus in monkeys cause linked deficits in smooth pursuit eye movements and adaptive modification of the VOR. *J. Neurophysiol.*, **87**, 912–924.

Richter, H.O., Costello, P., Sponheim, S.R., Lee, J.T. & Pardo, J.V. (2004) Functional neuroanatomy of the human near/far response to blur cues: eye-lens accommodation/vergence to point targets varying in depth. *Eur. J. Neurosci.*, **20**, 2722–2732.

Robinson, D.A. (1981) Control of eye movements. In: *Handbook of Physiology—The Nervous System II*, pp. 1275–1320. American Physiological Society, Bethesda, MD.

Roy, J.E. & Cullen, K.E. (2004) Dissociating self-generated from passively applied head motion: neural mechanisms in the vestibular nuclei. *J. Neurosci.*, **24**, 2102–2111.

Sander, T., Sprenger, A., Neumann, G., Machner, B., Gottschalk, S., Rambold, H. & Helmchen, C. (2009) Vergence deficits in patients with cerebellar lesions. *Brain*, **132**, 103–115.

Shelhamer, M. & Clendaniel, R.A. (2002) Context-specific adaptation of saccade gain. *Exp. Brain Res.*, **146**, 441–450.

Stone, L.S. & Lisberger, S.G. (1990) Visual responses of Purkinje cells in the cerebellar flocculus during smooth-pursuit eye movements in monkeys II. Complex spikes. *J. Neurophysiol.*, **63**, 1262–1275.

Suzuki, D.A. & Keller, E.L. (1988a) The role of the posterior vermis of monkey cerebellum in smooth-pursuit eye movement control. II. Target velocity related Purkinje cell activity. *J. Neurophysiol.*, **59**, 19–40.

Suzuki, D.A. & Keller, E.L. (1988b) The role of the posterior vermis of monkey cerebellum in smooth-pursuit eye movement control. I. Eye and head movement-related activity. *J. Neurophysiol.*, **59**, 1–18.

Suzuki, D.A., Noda, H. & Kase, M. (1981) Visual and pursuit eye movement-related activity in posterior vermis of monkey cerebellum. *J. Neurophysiol.*, **46**, 1120–1139.

Taig, E., Kuper, M., Theysohn, N., Timmann, D. & Donchin, O. (2012) Deficient use of visual information in estimating hand position in cerebellar patients. *J. Neurosci.*, **32**, 16274–16284.
Thach, W.T., Goodkin, H.P. & Keating, J.G. (1992) The cerebellum and the adaptive coordination of movement. *Ann. Rev. Neurosci.*, **15**, 403–443.
Timmann, D., Watts, S. & Hore, J. (1999) Failure of cerebellar patients to time finger opening precisely causes ball high–low inaccuracy in overarm throws. *J. Neurophysiol.*, **82**, 103–114.
Versino, M., Hurko, O. & Zee, D.S. (1996) Disorders of binocular control of eye movements in patients with cerebellar dysfunction. *Brain*, **119**, 1933–1950.
Wetts, R., Kalaska, J.F. & Smith, A.M. (1985) Cerebellar nuclear cell activity during antagonist cocontraction and reciprocal inhibition of forearm muscles. *J. Neurophysiol.*, **54**, 231–244.
Wolpert, D.M., Miall, R.C. & Kawato, M. (1998) Internal models in the cerebellum. *Trends Cogn. Sci.*, **2**, 338–347.
Zhang, H. & Gamlin, P.D.R. (1998) Neurons in the posterior interposed nucleus of the cerebellum related to vergence and accommodation. *J. Neurophysiol.*, **79**, 1255–1269.

## FURTHER READING

Barlow, J.S. (2002) *The Cerebellum and Adaptive Control*. Cambridge University Press, Cambridge.
Krauzlis, R.J. & Stone, L.S. (1999) Tracking with the mind's eye. *Trends Neurosci.*, **22**, 544–550.
Lisberger, S.G. (2009) Internal models of eye movement in the floccular complex of the monkey cerebellum. *Neuroscience*, **162**, 763–776.
Manni, E. & Petrosini, L. (2004) A century of cerebellar somatotopy: a debated representation. *Nat. Rev. Neurosci.*, **5**, 241–249.
Thach, W.T., Goodkin, H.P. & Keating, J.G. (1992) The cerebellum and the adaptive coordination of movement. *Ann. Rev. Neurosci.*, **15**, 403–443.
Wolpert, D.M., Miall, R.C. & Kawato, M. (1998) Internal models in the cerebellum. *Trends Cogn. Sci.*, **2**, 338–347.

# 9 Balance and Locomotion

An ibex runs straight down a cliff face to avoid a predator. For each step it must balance itself, locate the next foothold, and execute a motor plan. The less time it needs to complete all three tasks, the more likely it is to avoid becoming dinner. But the tasks all have to be done precisely; the ibex is, after all, on a cliff. It will come as no surprise that the cerebellum is involved in solving the problem of foot placement, just as it is in precision arm movements.

First, consider how an animal (or human) balances itself. When we say that someone has a "great sense of balance," what we usually mean is that that person can stay upright when he or she might be expected to fall over. The vestibular sense is important for balance, but balance is really motor performance that depends on several senses. It includes the coordinated action of the limbs that keeps us standing and small movements that deal with changes in the environment, such as the sway of a tightrope.

Some balance functions are carried out by spinal reflexes alone. Somatosensory and proprioceptive sense organs in the skin, muscles, and joints can detect changes in posture for feedback control, or a sudden problem with the supporting surface. As we walk, proprioceptors monitor forces that act on the feet and joints, and regulate muscle activity in the legs by way of reflex pathways in the spinal cord. Increasing the weight on the legs, for example, is immediately detected by proprioception and compensated by increased tension in flexor muscles (Prochazka *et al.* 1997). Spinal reflexes can accommodate the changes in load on the muscles that occur during the step cycle, and that are caused by changes in the slope of the substrate during walking (Hatz *et al.* 2012).

In addition to spinal reflexes, balance relies heavily on the vestibular sense organs in the labyrinth of the inner ear. The vestibular sense acts as a second source of muscle tone, provided directly to the motoneurons, and provides immediate feedback if we start to fall. We are not usually even aware of our vestibular sense, but we certainly notice if we lose it. Occasionally, this happens to patients who have aminoglycoside

*The Cerebellum: Learning Movement, Language, and Social Skills*, First Edition. Dianne M. Broussard.

poisoning.[1] These patients are still able to walk, but they need visual feedback to stabilize their bodies. At night, they need to put a hand on a wall or furniture even to *stand* without falling. They also have severe problems with gaze instability (because they have lost the eye and neck vestibular reflexes).

Locomotion is especially complex for humans, who balance on two feet. Standing requires continuous processing of visual, vestibular, somatosensory and proprioceptive signals, and continuous adjustment of the muscles of the legs, trunk, neck, even the eyes (e.g., we need to stabilize gaze in order to stand on one foot). Maintaining balance requires integration of the various sensory inputs in a feedback control system. The cerebellum processes most of these sensory signals, and coordinates the muscles for balance. During stepping, balance control is more challenging. For example, during standing, extensors are activated to straighten up after a lateral push, but during walking, hip abductors and adductors are also needed (Karayannidou *et al.* 2009).

The contribution of the cerebellum to stance and gait is responsible for the best-known sign of cerebellar damage: locomotor ataxia, also known as "cerebellar ataxia."

## CEREBELLAR ATAXIA

Many cerebellar patients have such a distinctive walk that an amateur can spot them. They have a broad-based gait (the feet are farther apart than normal). Steps are short and unequal. The patient also seems unsteady and is more likely to fall than a normal person. In humans, the forebrain is also necessary for normal walking. In quadrupeds, the forebrain is not even necessary for walking, but the various regions of the cerebellum all play a part. For a start, the deep cerebellar nuclei generate a steady stream of action potentials that provide excitatory drive to the motoneurons. The drive from the cerebellum is a major contributor to muscle tone. Without muscle tone, we are not going anywhere. Also, the anterior lobe of the cerebellum is directly interconnected with the pattern generator for locomotion. The cerebellum controls coordination across joints and feedback control, or servocontrol, of the legs. In alcoholic cerebellar atrophy, which primarily affects the anterior lobe, gait is uncoordinated and patients cannot hop (Thach & Bastian 2004). They also have intention tremor in the legs, and sometimes sway while standing still.

The cerebellum plays a key role in using multiple senses for balance. We use whatever sensory information is available to balance while we

[1] Aminoglycoside antibiotics such as gentamicin can damage the hair cells of the vestibular labyrinth. The best-known case of this is "J. C." (JC 1952).

are walking, and in fact we can switch rapidly among senses, choosing whichever appears to be more reliable at the moment (Fetsch *et al.* 2009). Proprioceptive information about joint rotation and somatosensory input about the surface we are walking or running on are essential. In fact, if the head is artificially stabilized, proprioceptive feedback suffices for standing, without any visual or vestibular input (Musienko *et al.* 2012). But stabilizing the head is one of the major problems faced by the balance system. Vestibular and/or visual information is necessary under normal conditions.

Using the different senses brings in different regions of the cerebellum. Roughly speaking, somatosensory and proprioceptive information flow into the anterior lobes; vestibular input to the vestibulocerebellum, and visual input to the vestibulocerebellum and the posterior lobes. Accordingly, ataxia may have fundamentally different causes in different cases.

## *The lurcher: a mutation affecting locomotion*

Like humans with cerebellar damage, lurcher mice have locomotor ataxia. The lurcher mutant[2] is a gain-of-function mutation of the $\delta 2$ glutamate receptor. It results in the loss of nearly all Purkinje cells (P-cells), and most granule cells as well, during maturation. Many of the lurcher deficits can be attributed to poor coordination. First, they have trouble making their way rapidly along the bottom of a coat hanger (amazingly, normal mice have no problem with this) (Lalonde *et al.* 1992). Second, strength is low, as would be expected if co-contraction is impaired. Lurchers have difficulty climbing on a vertical grid and walking on a perforated board, suggesting that foot placement is impaired. They also seem to have a balance defecit, because they adopt a strategy of clinging rather than walking on a balance beam and are not able to stay on the beam as long as normal mice. They fail miserably at a test with an unstable platform, which requires them to make postural adjustments (Hilber & Caston 2001).

The rotorod test requires running at the correct speed but also tests the vestibular sense, since mice must constantly monitor departures from vertical. The poor performance on the rotorod in lurcher (Hilber & Caston 2001) may be related to a learning deficit, but could also be related to timing and/or balance.

There are several other mutations that cause P-cell degeneration (Lalonde & Strazielle 2007), and similar dysfunction is seen in many of these. At least one other mutant lacking functional $\delta 2$ receptors walks less, shows moderate locomotor ataxia, does poorly on the rotorod, and

---

[2]The lurcher was a spontaneous mutation that was first reported by R. J. S. Phillips (Phillips 1960).

has more "foot slips" when walking on a 2-cm-wide beam (Kashiwabuchi *et al.* 1995).

## SIGNALS FROM THE INNER EAR

In order to balance during bipedal walking, we must align the body with the gravity vector (i.e., vertical). To do this, we monitor the environment using our visual and vestibular senses. If light is available, we use visual cues for proper body alignment; these can include walls, furniture, and other people. But whether or not there is light, we need to detect any tilt of the head from vertical and adjust the muscle tone in the neck, trunk, and legs. This is where the vestibular labyrinth is essential. The static sense of balance—staying upright—is mediated mostly by the otolith organs of the utricle and saccule, which respond to tilt. Dynamic balance and the sense of motion are mediated both by the semicircular canals, which respond to rotation, and by the response of the otolith organs to translation.

Normally, when we begin to tilt, the change in the vestibular signal brings about a compensatory change in muscle tone that prevents a fall. Some neurons in Deiters' nucleus (also called the lateral vestibular nucleus) project directly to motoneurons that in turn innervate the trunk and leg muscles. Because the postural adjustments work by changes in muscle tone, damage to the otoliths causes disorders of muscle tone and falling. The continuous signal from the semicircular canals is also a fairly high discharge rate (around 50 action potentials per second). Losing this signal on one side can cause falling, circling or head nodding, and nystagmus. Signals from the semicircular canals contribute to the compensatory response to tilt.

Proprioceptive information, visual information, and vestibular information each originate in frames of reference that depend on the sense organs. For example, proprioception has a frame of reference that is similar to the "motor coordinates" that were explained in Chapter 8, while semicircular-canal signals are encoded with respect to the head, in head coordinates,[3] and visual signals are in retinal coordinates. But for balance, a body-centered frame of reference is useful. In order to get to a body coordinate frame, we must combine proprioceptive information about neck position on the body with vestibular information about head position in space, to derive the body's position in space. Like the conversion from

[3]An example of head coordinates is the definition of a nod. A nod is defined in head coordinates as a movement in the sagittal plane. The spatial plane of a nod might be either vertical (if standing) or horizontal (if lying down). The key here is that regardless of the spatial plane, the movement remains the same with respect to the head.

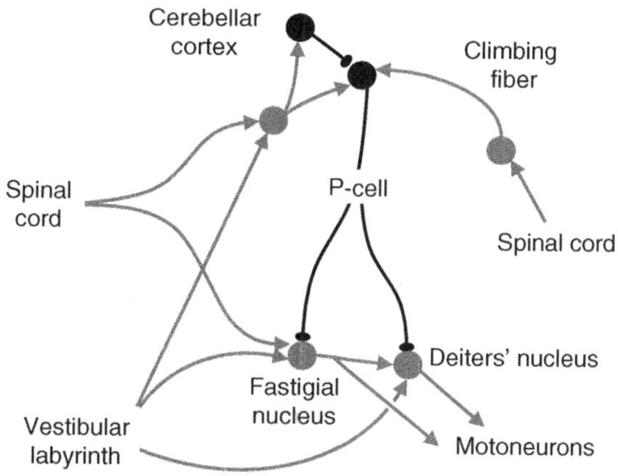


**Figure 9.1** A simplified diagram of the control of balance by vestibular and proprioceptive signals, by way of the cerebellum. The vestibular system provides input to the anterior lobe and the vestibulocerebellum, and also directly to the fastigial and Deiters' nuclei. The spinal cord provides proprioceptive and somatosensory input to the cerebellar cortex, deep nuclei, and inferior olive, and also provides the locomotor rhythm. Vestibular, proprioceptive and somatosensory input converge both in the cerebellar cortex and in the fastigial nucleus. Both the fastigial nucleus and Deiters' nucleus project to motoneurons. Several connections have been omitted.

sensory to motor coordinates for coordination of arm and hand movements, this complex calculation seems to be a perfect task for the cerebellum.

The cerebellar network controlling balance includes the vestibulocerebellum, anterior lobe, and fastigial nucleus. The close relationship of these components to the vestibular and proprioceptive sensory pathways is illustrated in Figure 9.1.

## *Balance and the vestibulocerebellum*

Lesions that are restricted to one side of the vestibulocerebellum—including the flocculus, ventral paraflocculus, lobules IX and X of the vermis, the fastigial nucleus, or certain parts of the inferior olive—cause a set of signs very much like those of damage to the inner ear. A lesion of lobule X that includes the flocculus causes nystagmus. Human subjects are completely unable to perform tandem (heel-to-toe) walking after stroke damage to lobule X, and often fall toward the lesioned side (Thach & Bastian 2004).

In a classic study on monkeys by Botterell and Fulton, after a lesion of lobules IX and X of the vermis (or of the fastigial nuclei on both sides), the subjects nodded their heads constantly; they had a postural sway and could not stay upright without support; the limbs tended to remain either flexed or extended, sometimes splaying outward (so that the hindquarters collapsed), and movements tended to be too *fast* (more about this later) (Botterell & Fulton 1938a). Over several months, the monkeys partly recovered. Using the remaining cerebellum and the forebrain, they were able to improve, although they never returned to normal. These results illustrate how intimately involved the cerebellum is in the brain's network for controlling balance. Without the computations that the cerebellum does, it all falls apart, even though the spinal reflexes and the inner ear are still intact.

## Balance and the fastigial nucleus

Balance and locomotion functions are carried out by the rostral part of the fastigial nucleus. The rostral fastigial nucleus receives input from vermal lobules IX and X and projects to the motoneurons and the vestibular nuclei (Figure 9.1). As we saw in Chapter 8, the fastigial nucleus has a role in coordinating the eye muscles for vergence, and in stopping saccades, but these oculomotor functions are carried out by the caudal fastigial nucleus.

During walking, as during standing, muscle tone must be regulated; for example, extensor tone that is too high will shorten the stride, and if it is much too high the legs become rigid. In Botterell's monkeys, if lobule X was destroyed, muscle tone was too high, but if the fastigial nucleus was damaged, muscle tone was too low. This is because the fastigial nucleus is inhibited by P-cells in lobule X, and the fastigial nucleus in turn provides excitatory input to the vestibular nuclei. The vestibular nuclei regulate muscle tone at the motoneuron level. The anterior lobe of the cerebellum contributes to tone by modulating activity, both in the fastigial nucleus and in the Deiters' nucleus. The vestibular labyrinth provides much of the excitatory input to the muscles, and this is supplemented by inhibitory input from the vestibulocerebellum that converges on the same downstream neurons.

In order to maintain posture, the vestibular reflexes need information about what the posture is exactly. For different body positions, different muscles must be activated. This requirement is met using proprioceptive signals, which interact with vestibular signals at the fastigial nuclei. This interaction transposes the vestibular signals in the fastigial nuclei into the body-centered reference frame, so that in the fastigial nucleus, body tilt rather than merely head tilt is now represented. The calculation appears to be done in the fastigial nucleus itself, possibly using converging information from the vestibulocerebellum and the anterior lobe (Kleine *et al.* 2004). The inaccurate voluntary movements and movements that are "too fast" after lesions of the vestibulocerebellum could be related to an inability to use proprioceptive feedback.

Locomotion depends on the nucleus interpositus as well as the fastigial nucleus. The interpositus is involved in proprioceptive feedback and error correction for limb movements, and like the fastigial nucleus, it regulates muscle tone. In monkeys and cats, damage to the nucleus interpositus causes extensors to have a higher tone than flexors, so that the foot scrapes along the ground and may get folded over and dragged. In contrast, inactivation of the overlying cortex in lobule V (anterior lobe) causes the opposite situation: Excessive flexion of the joints of the forelimb tends to prolong the swing phase of walking (Udo *et al.* 1980). Similar problems with foot dragging or excessive flexion were seen in some of Holmes' patients (Holmes 1917).

The fastigial nucleus receives a small amount of visual input, as do the vestibular nuclei. Visual input to the fastigial nucleus arises not from the cortical/pontine pathway but from the retina by way of a shortcut, the accessory optic pathway, which is entirely within the brainstem. A minority of the fastigial neurons respond to visual stimuli, especially rotation of the visual field in the roll or pitch planes that simulates tilt (Bryan & Angelaki 2009).

## Balance and the inferior olive

The sensory responsiveness of the vestibulocerebellum depends not only on its input from the vestibular labyrinth, but also on climbing-fiber function, which closes the cerebellar cortical-fastigial-inferior olivary loop. Inactivation of the olive on one side has an immediate effect on balance, because climbing-fiber inputs directly affect the P-cell simple-spike output; loss of the complex spikes causes simple-spike discharge to increase to a high, regular rate.

Particular regions of the inferior olive are necessary for the responses of P-cells in lobules IX and X to vestibular input (Barmack & Yakhnitsa 2011). These regions are the caudal medial accessory olive (cMAO) and the beta-nucleus. Inactivating the cMAO in a cat causes the animal to fall away from the inactivated side. The subject is usually unable to walk unless it can use a wall for support. If it is able to walk, the path deviates away from the injected side. Foot placement and reach to grasp are not affected, suggesting that coordination is not a factor. Inactivation of the beta-nucleus results in locomotor ataxia, with small, uneven steps due to short swing phases—possibly reflecting increased muscle tone—and swaying away from the injected side. The cats also lose their appetites, suggesting a vestibular sensory deficit (Horn *et al.* 2010).[4]

The rostral medial accessory olive (rMAO) projects to the cortex overlying the nucleus interpositus. In contrast to the vestibular parts of the inferior olive, inactivation of the rMAO produces decreased muscle tone

[4]Unilateral vestibular loss causes vertigo, an unpleasant spinning sensation that in turn causes nausea.

that leads to locomotor ataxia. The contralateral limbs are weak, causing the cat to limp, and some animals develop a "waddle," that is, a very broad-based gait with sway and foot dragging. Trunk muscle tone also seemed to be affected. Instead of walking back and forth in a straight line, the animal follows a figure eight trajectory (Horn *et al.* 2010). rMAO neurons respond to muscle squeeze, suggesting that proprioceptors provide the sensory input that helps the rMAO module to regulate muscle tone.

## COORDINATING LOCOMOTION

Although good balance is essential for normal walking, it is not all that is needed. We also have to start the locomotor rhythm and tell the legs what to do. Central pattern generators in the spinal cord generate a rhythm for walking and provide alternating flexor and extensor activity in each group, which is the motor command for walking. They activate muscle groups that act synergistically, such as extensors in the right arm and the left leg. Fastigial neurons respond to this rhythm; they fire rhythmically during fictive locomotion and also during fictive scratching. In fictive movements, there is no actual movement, indicating that the response is not due to sensory feedback signals; instead, the fastigial nucleus responds to input from the central pattern generators (Antziverofa *et al.* 1980). Deiters' nucleus also participates in the locomotor rhythm (Orlovsky 1972).

Depending on the surface, foot placement during walking may require planned movements, and they can be guided by vision and gaze feedback using the vestibulocerebellum and/or the cerebellar posterior lobes. Processed proprioceptive signals, possibly arising from the supplementary motor areas of the cerebral cortex, are also used for motor planning by the posterior lobes (Taig *et al.* 2012).

Damage to the posterior-lobe hemispheres or to the dentate nucleus affects locomotor behavior and certain characteristics of gait. Botterell and Fulton found that removal of the cortex of the cerebellar hemisphere did not cause ataxia. Instead, the monkey would repeatedly gallop right into a solid door; it either could not see the door, or could not stop. Leaping was abnormally high and very noticeable. This is wildly different from the classic cerebellar signs. Lesions of the dentate nucleus caused a different syndrome; they were especially likely to cause tremor, suggesting a problem with feedback. But high leaping ("grasshopper leaping") was also seen with dentate nuclear lesions (Botterell & Fulton 1938b).

Neurons in the cerebellum respond vigorously during visually-guided walking. When walking on a ladder, which requires careful foot placement, P-cells, interpositus neurons, and dentate neurons in cats all fire rhythmically in phase with stepping, with most increasing their discharge during the swing phase. These same neurons also respond either to visual stimuli and/or to saccades (Marple-Horvat & Criado 1999). Many P-cells in the hemisphere and underlying dentate neurons also show visual responses (Marple-Horvat *et al.* 1998). These results indicate that information about visual targets may be explicitly represented in the posterior lobe and dentate nuclei.

## *Complex-spike signals during locomotion*

We have seen that the inferior olive is necessary for balance. But what, exactly, do the climbing fibers contribute to locomotion? In one theory, climbing fibers fire more during situations where walking becomes challenging and falls are likely. Their discharges may encode the sensory signal—whether somatosensory, vestibular, or visual—that represents the difficulty. In these situations, corrective foot movements must be quick, precise, and accurate and inter-limb coordination is also required.

Climbing fibers do respond to some challenges. A cat walks on a horizontal ladder with particular rungs that suddenly "give." In this task, P-cells in lobule V of the vermis generate complex spikes just after the perturbation (Andersson & Armstrong 1987; Yanagihara & Udo 1994). The cat is able to correct its foot placement quickly and avoid falling. In a similar task where a rung is suddenly elevated, mice with impaired excitatory transmission to P-cells make more missteps than normal mice (Saab *et al.* 2012). The gamma type of protein kinase C (PKC$\gamma$) mutant mice retain multiple climbing-fiber innervation of P-cells,[5] and also show poor performance on the rotorod. The impairment is quite dramatic and is evident on the first trial. The mutant mice cannot stay on the rotating rod more than a few seconds, nor can they learn to stay on a thinner rod even when it is stationary. Interestingly, the mutant mice can learn eyeblink timing and can learn to stay on the stationary rotorod (Chen *et al.* 1995). These observations are consistent with signaling of sudden changes in the environment by climbing fibers.

Normal activity in climbing fibers also seems to be important for coordinating the toes as well as the limbs. During inactivation of the rostral dorsal accessory olive (rDAO) in cats, the paw tends to be closer to the floor than normal during the swing phase. The front or hind paw is affected depending on the affected region of rDAO, which is somatotopically organized. The toes also fail to participate in a reach-to-grasp task. The deficit from rDAO inactivation is similar to that after lesions of the anterior interpositus or the overlying paravermis, which receive input from this part of the olive. It is also similar to lesions of the red nucleus

[5]In wild-type mice, multiple innervation is present during development, but multiple innervation is rare in adult wild-type mice.

(a projection target of the interpositus). This all suggests that toe coordination during walking is controlled by this particular cerebellar module. Inactivating the climbing fibers may effectively inactivate the entire module (Horn *et al.* 2010).

## NAVIGATION

In addition to staying upright and coordinating our steps, we need to steer a path to where we want to go, or at the very least, steer a path away from places we do not want to go. Both vestibular and visual information, integrated with information about gaze direction, are used for steering.

Although there are exceptions, people usually do look where they are going. Gaze direction is generally the direction that we want to go in; therefore we need to represent gaze direction in the brain in such a way that we can organize our steps toward it. The first task is to point both foveae at the goal, using a saccade. If we decide to go toward the target, we generally point our noses at it as well. If the head is directed at the target, gaze direction can then be calculated using information from neck proprioceptors combined with eye position information. All of these modes of information are represented by the cerebellum.

Our actual motion in space can be derived from visual and vestibular signals. Visual information provides cues to our speed and direction, based on optic flow. Optic flow is motion on the retina of the image of the visual scene. For example, if we move forward toward a point that we are looking at, the motion of the scene is outward from the fovea. The vestibulocerebellum receives information about optic flow from the entire visual field on both its climbing-fiber and mossy-fiber input pathways. The visual responses of P-cells in lobules IX and X indicate whether we are moving forward, backward, up, or down. The direction of optic flow is consistent within a cerebellar zone, but changes across zones (Graham & Wylie 2012).

The body-centered reference frame is essential for posture and also for navigation. The conversion of vestibular information to the body-centered reference frame is done by combining proprioceptive information about neck position and movement on the body with vestibular information about head position and movement. This might be done by the anterior lobe with the other members of its team, the fastigial nucleus and the inferior olive. Some fastigial neurons explicitly encode the body-rotation signal in their discharge patterns (Brooks & Cullen 2009). Visual information and proprioceptive information also need to be transposed to the correct frame of reference. Information about body movement and position is useful not only for planning locomotion and postural reflex adjustments, but also for planning accurate throwing and other limb movements.

## Calculating tilt and translation

Another way of navigating, which does not require us to maintain continuous fixation on a goal, is to develop a representation of both the goal and the body in space, that is, in a spatial frame of reference. For example, we can get up at night and walk around in the dark. In such a situation we may use a form of dead reckoning, that is, keeping track of the movements we have made toward a goal whose location we know in spatial coordinates. We can keep track of rotations and translations of our body in space using a processed form of vestibular signals that is integrated with neck proprioceptive information.

As already mentioned, rotation, tilt, and translation are all signaled by the vestibular labyrinth. The translation signal (linear motion through space) is potentially useful in keeping track of our movements, particularly in the dark. However, there is a problem: the vestibular sensory organs respond in exactly the same way to translation and tilt. If I tilt my head backward, gravity displaces the otoliths[6] in exactly the same way that inertia would displace them if I had moved forward. For body-stabilizing reflexes, this might be acceptable, since we need to increase extensor tone in the back leg in both situations. But for dead reckoning, it is essential to be able to separate tilt from translation. Fortunately, the vestibulocerebellum does this for us; in fact, this might have been one of the very first functions that was performed by the cerebellum in our primitive vertebrate ancestors.

Work in the laboratory of Dora Angelaki has outlined the process of discriminating tilt from translation. P-cells in lobules IX and X receive input from the semicircular canals when the head rotates *vertically* (that is, nodding from front to back, or tilting from side to side). When there is a (ambiguous) signal from the otolith organ, the vestibulocerebellum uses the input from the semicircular canals to decide whether the head has rotated or not. If it has, then the otolith signal is interpreted as tilt. If not, then the otolith signal is interpreted as linear motion (Angelaki & Cullen 2008).

The dorsomedial cell column of the inferior olive receives input from the otoliths and project to lobules IX and X of the vestibulocerebellum. Inactivation of this region causes no deficit until the head is tilted, but when it is, the animal rotates its head rhythmically in an accelerating rhythm, followed by an immediate collapsing of the limbs (Horn *et al.* 2010). Without the computations performed by the cerebellum, the cat cannot stabilize its head; driven by inaccurate sensory information, the reflexes generate too much tilt. This strange result provides evidence that translation and roll tilt cannot be separated without the vestibulocerebellum.

---

[6]The otoliths are microscopic crystals that balance on top of cilia inside two tiny organs, the utricle and the saccule, that are located in the inner ear.

Further, it emphasizes that the so-called "vestibulospinal" reflexes, in fact, need the cerebellum to operate correctly.

Surprisingly, the P-cells in IX and X do not overtly respond to tilt; instead they respond selectively to linear motion. This means that the cerebellar cortical circuitry uses the canal signal to calculate a pure linear motion signal. The canal signal that these neurons receive is completely invisible in their responses; it has been used to cancel the tilt input from the otolith organs (Yakusheva *et al.* 2007). This linear motion signal is already in spatial coordinates, that is, it is in the correct frame of reference for navigation using dead reckoning.

## CONCLUSIONS

The cerebellum does a large number of essential computations to support locomotion. The anterior lobes are important in the integration of vestibular and proprioceptive information so that we can use it to guide our movements and provide feedback about errors. The posterior lobes are involved in planning of foot placement during locomotion in demanding situations, and also in guiding our movements, and orienting our bodies correctly, using visual signals. The vestibulocerebellum processes information from the inner ear, proprioceptors, and the retina to tell us in what direction, and how fast, we are moving, calculates head orientation so that we can stay upright, and precisely regulates muscle tone so that we can move quickly without tripping. The deep cerebellar nuclei contribute additional processing and integrate information from the cerebellar cortex before sending it to motor regions of the cerebral cortex and to the spinal cord.

## REFERENCES


Andersson, G. & Armstrong, D.M. (1987) Complex spikes in Purkinje cells in the lateral vermis (b zone) of the cat cerebellum during locomotion. *J. Physiol.*, **385**, 107–134.

Angelaki, D.E. & Cullen, K.E. (2008) Vestibular system: the many facets of a multimodal sense. *Ann. Rev. Neurosci.*, **31**, 125–150.

Antziverofa, L.I., Arshavsky, Y.I., Orlovsky, G.N. & Pavlova, G.A. (1980) Activity of neurons of cerebellar nuclei during fictitious scratch reflex in the cat. I. Fastigial nucleus. *Brain Res.*, **200**, 239–248.

Barmack, N.H. & Yakhnitsa, V. (2011) Microlesions of the inferior olive reduce vestibular modulation of Purkinje cell complex and simple spikes in mouse cerebellum. *J. Neurosci.*, **31**, 9824–9835.

Botterell, E.H. & Fulton, J.F. (1938a) Functional localization in the cerebellum of primates. II. Lesions of midline structures (vermis) and deep nuclei. *J. Comp. Neurol.*, **69**, 47–62.

Botterell, E.H. & Fulton, J.F. (1938b) Functional localization in the cerebellum of primates. III. Lesions of hemispheres (neocerebellum). *J. Comp. Neurol.*, **69**, 63–87.

Brooks, J.X. & Cullen, K.E. (2009) Multimodal integration in rostral fastigial nucleus provides an estimate of body movement. *J. Neurosci.*, **29**, 10499–10511.

Bryan, A.S. & Angelaki, D.E. (2009) Optokinetic and vestibular responsiveness in the macaque rostral vestibular and fastigial nuclei. *J. Neurophysiol.*, **101**, 714–720.

Chen, C., Kano, M., Abeliovich, A., Chen, L., Bao, S., Kim, J.J., Hashimoto, K., Thompson, R.F. & Tonegawa, S. (1995) Impaired motor coordination correlates with persistent multiple climbing fiber innervation in PKCγ mutant mice. *Cell*, **83**, 1233–1242.

Fetsch, C.R., Turner, A.H., DeAngelis, G.C. & Angelaki, D.E. (2009) Dynamic reweighting of visual and vestibular cues during self-motion perception. *J. Neurosci.*, **29**, 15601–15612.

Graham, D.J. & Wylie, D.R. (2012) Zebrin-immunopositive and -immunonegative stripe pairs represent functional units in the pigeon vestibulocerebellum. *J. Neurosci.*, **32**, 12769–12779.

Hatz, K., Mombaur, K. & Donelan, J.M. (2012) Control of ankle extensor activity in walking cats. *J. Neurophysiol.*, **108**, 2785–2793.

Hilber, P. & Caston, J. (2001) Motor skills and motor learning in Lurcher mutant mice during aging. *Neuroscience*, **102**, 615–623.

Holmes, G. (1917) The symptoms of acute cerebellar injuries due to gunshot injuries. *Brain*, **40**, 461–535.

Horn, K.M., Pong, M. & Gibson, A.R. (2010) Functional relations of cerebellar modules of the cat. *J. Neurosci.*, **30**, 9411–9423.

JC (1952) Living without a balancing mechanism. *N. Engl. J. Med.*, **246**, 458–460.

Karayannidou, A., Zelenin, P.V., Orlovsky, G.N., Sirota, M.G. & Beloozerova, I.N. (2009) Maintenance of lateral stability during standing and walking in the cat. *J. Neurophysiol.*, **101**, 8–19.

Kashiwabuchi, N., Ikeda, K., Araki, K., Hirano, T., Shibuki, K., Takayama, C., Inoue, Y., Kutsuwada, T., Yagi, T., Kang, Y., Aizawa, S. & Mishina, M. (1995) Impairment of motor coordination, Purkinje cell synapse formation, and cerebellar long-term depression in GluRd2 mutant mice. *Cell*, **81**, 245–252.

Kleine, J.F., Guan, Y., Kipiani, E., Glonti, L., Hoshi, M. & Buttner, U. (2004) Trunk position influences vestibular responses of fastigial nucleus neurons in the alert monkey. *J. Neurophysiol.*, **91**, 2090–2100.

Lalonde, R. & Strazielle, C. (2007) Spontaneous and induced mouse mutations with cerebellar dysfunctions: behavior and neurochemistry. *Brain Res.*, **1140**, 51–74.

Lalonde, R., Botez, M.I., Joyal, C.C. & Caumartin, M. (1992) Motor abnormalities in Lurcher mutant mice. *Physiol. Behav.*, **51**, 523–525.

Marple-Horvat, D.E. & Criado, J.M. (1999) Rhythmic neuronal activity in the lateral cerebellum of the cat during visually guided stepping. *J. Physiol.*, **518**, 595–603.

Marple-Horvat, D.E., Criado, J.M. & Armstrong, D.M. (1998) Neuronal activity in the lateral cerebellum of the cat related to visual stimuli at rest, visually guided step modification, and saccadic eye movements. *J. Physiol.*, **506**, 489–514.

Musienko, P., Courtine, G., Tibbs, J.E., Kilimnik, V., Savochin, A., Garfinkel, A., Roy, R.R., Edgerton, R. & Gerasimenko, Y. (2012) Somatosensory control of balance during locomotion in decerebrated cat. *J. Neurophysiol.*, **107**, 2072–2082.

Orlovsky, G.N. (1972) Activity of vestibulospinal neurons during locomotion. *Brain Res.*, **46**, 85–98.

Phillips, R.J.S. (1960) "Lurcher", a new gene in linkage group XI of the house mouse. *J. Genet.*, **57**, 35–42.

Prochazka, A., Gillard, D. & Bennett, D.J. (1997) Positive force feedback control of muscles. *J. Neurophysiol.*, **77**, 3226–3236.

Saab, A.S., Neumeyer, A., Jahn, H.M., Cupido, A., Simek, A.A.M., Boele, H.-J., Scheller, A., Le Meur, K., Gotz, M., Monyer, H., Sprengel, R., Rubio, M.E., Dietmer, J.W., De Zeeuw, C.I. & Kirchhoff, F. (2012) Bergmann glial AMPA receptors are required for fine motor coordination. *Science*, **337**, 749–753.

Taig, E., Kuper, M., Theysohn, N., Timmann, D. & Donchin, O. (2012) Deficient use of visual information in estimating hand position in cerebellar patients. *J. Neurosci.*, **32**, 16274–16284.

Thach, W.T. & Bastian, A.J. (2004) Role of the cerebellum in the control and adaptation of gait in health and disease. *Prog. Brain Res.*, **143**, 353–366.

Udo, M., Matsukawa, K., Kamei, H. & Oda, Y. (1980) Cerebellar control of locomotion: effects of cooling cerebellar cortex in high decerebrate and awake walking cats. *J. Neurophysiol.*, **44**, 119–134.

Yakusheva, T.A., Shaikh, A.G., Green, A.M., Blazquez, P.M., Dickman, J.D. & Angelaki, D.E. (2007) Purkinje cells in posterior cerebellar vermis encode motion in an inertial reference frame. *Neuron*, **54**, 973–985.

Yanagihara, D. & Udo, M. (1994) Climbing fiber responses in cerebellar vermal Purkinje cells during perturbed locomotion in decerebrate cats. *Neurosci. Res.*, **19**, 245–248.

## FURTHER READING

Thach, W.T. & Bastian, A.J. (2004) Role of the cerebellum in the control and adaptation of gait in health and disease. *Prog. in Brain Res.*, **143**, 353–366.

Fitzpatrick, R.C., Butler, J.E. & Day, B.L. (2006) Resolving head rotation for human bipedalism. *Curr. Biol.*, **16**, 1509–1514.

Angelaki, D.E. & Cullen, K.E. (2008) Vestibular system: The many facets of a multimodal sense. *Ann. Rev. Neurosci.*, **31**, 125–150.

# 10 Timing

Timing is everything, especially when an animal is hunting (or being hunted), or dealing with another challenging situation. Well-timed starts and stops make some movements precisely reproducible, giving them terrific accuracy. Everyone has their favorite examples: virtuoso piano playing, say, or synchronized diving, where movements of two divers are not only timed but synchronized within a few tenths of a second. But we do not have to be skilled musicians or athletes to be good at timing; running down a flight of stairs is an amazing, but quite ordinary, feat of precision timing. How do we do this sort of thing?

It is very clear that the cerebellum is important for motor timing. For any brisk movement, contractions of agonist and antagonist muscles, and muscles controlling different joints, must be accurately timed. Gordon Holmes' cerebellar patients exhibited delays both in starting and in stopping movement. They also had difficulty with movement sequences that required rapid alternating contraction of agonist and antagonist muscle groups (Holmes 1917); this condition has the wonderful name of dysdiadochokinesia. Cerebellar damage can also cause rapid orienting movements to overshoot because they fail to stop on time.

Cerebellar patients show a deficit of interval timing, revealed by a task that simply requires the subjects to tap with a finger in a fast rhythm (Ivry & Keele 1989). Mistimed finger movements can cause other problems. Inaccurate throwing in cerebellar patients, for example, is partly due to inaccurate timing of finger opening (Timmann *et al.* 1999). Timing is part of most of the simple motor skills that were described in Chapters 5–9, including the nictitating-membrane response (NMR), eye movements, pointing, throwing, coordination, balance, and locomotion.

Timing also has other uses besides motor control. It is important for auditory processing of speech sounds. Discrimination of auditory intervals, like the generation of timed movements, is impaired in cerebellar patients who also show deficits of motor interval timing while in contrast, there is not always a correlation between the ability to generate timed rhythmic movements and the ability to generate precisely controlled force (Ivry & Keele 1989)). Perhaps, timing is a separate cerebellar function that can be used by both motor and sensory systems. In this chapter, I will return

*The Cerebellum: Learning Movement, Language, and Social Skills*, First Edition. Dianne M. Broussard.

to some of the tasks that the cerebellum performs, and describe some possible cerebellar mechanisms that could carry out timing functions for each task.

## TIMING USING DISCHARGE RATES

A relatively simple example of cerebellar timing is the conditioned NMR (Chapter 5). The conditioned NMR is triggered by a tone, and its occurrence is timed by the cerebellar cortex. Some of the changes that occur in the cerebellar cortex during learning must make the correct timing happen, because the timing of the learned NMR is expressed in the discharge patterns of Purkinje cells (P-cells). The correct time of the NMR is represented as a pause in the steady stream of simple spikes (Hesslow & Ivarsson 1994). This learned pause in P-cell firing disinhibits the interpositus, generating a timed blink. Timing information is thought to remain stored in the cortical circuitry, while additional synaptic plasticity in the interpositus enables triggering of the NMR by a tone.

A large, simulated neural network with the characteristics of the cerebellar cortex, developed by Dean Buonomano and Michael Mauk, can learn timing that is accurate enough for the NMR (Buonomano & Mauk 1994). In the model, the conditioned stimulus is represented by a particular pattern of activity in the population of mossy-fiber inputs, which results in a particular pattern of parallel-fiber inputs to P-cells. The key to timing the response is that parallel-fiber activity reaches a peak at different times after the tone in different populations of granule cells. Buonomano and Mauk proposed that Golgi cells are responsible for generating the different peaks. Golgi cells inhibit some granule cells, and the discharges of those cells will have a delayed peak. Given the variation in Golgi-cell inhibition, a subset of parallel fibers will always become active around the time of the climbing-fiber spike, which represents the time of the unconditioned stimulus (the best time to generate a blink). Those "optimal" parallel fibers become the ones whose input is decreased by LTD. As a result, in the model, P-cells fire less during the optimal time of the blink. As a result of decreased P-cell firing, deep nuclear neurons increase their discharge in turn, and the blink occurs. Adding cerebellar LTP in a later version of the model permitted P-cells to learn to fire more during the early part of the CS, preventing early responses in deep nuclear neurons (Medina *et al.* 2000).

In Mauk's models, timing of the NMR is encoded as the "correct" subset of granule cells, in a "place code."[1] It works like this: the activity of

[1]In a place code, a variable is represented as a particular active group of neurons, as a town is represented on a map. There are many examples of place codes in sensory systems; features of sounds such as frequency and visual objects such as color are encoded in the sensory cortex using place codes.

each group of granule cells can be represented as a population vector, which has as many dimensions as there are neurons in the group. As firing rates change the vector changes in a nonrepetitive way, encoding time with high precision. By depressing the parallel fiber–P cell synapses selectively for those granule cells that contribute to the vector during the correct time window, timing of the P-cell pause can be optimized. The cerebellar clock, in the models from Mauk's group, consists of the cortical circuitry with appropriate synaptic weights to produce the appropriate delays. It does not require any specialized structure, and notably, it does not require complex spikes, except to bring about synaptic plasticity. In fact, Mauk's models of the cerebellum do not have rhythmic activity. They perform a timing function, but in a fundamentally different way from an actual clock.

## *Linear filters*

Models like Mauk's are not constrained to be linear. However, some simpler models are, and many of the concepts related to timing that we will explore in this chapter rely on linear filters. So why can't cerebellar filtering be done entirely by linear filters? In the eyeblink task, the job of the cerebellar cortex is to generate a fixed delay between the tone and the blink. A pure delay is a discontinuous process: signaling occurs, we wait for the prescribed amount of time, and then a response occurs. A pure delay also lasts a certain number of milliseconds, no matter what the signal frequency happens to be.

A linear filter is a simple device that may be used by the brain in many situations. It generates an output that is always proportional to its input. Here is the important point: there is no way to generate a pure delay with a linear filter. This is because a linear filter depends on a varying input signal (the discharge rate). If the signal stops, a linear filter has no way of making something happen later. Linear filters can generate delays in some situations; these are not pure delays, but are based on the frequency of a continuous signal. Still, they are useful, because under normal conditions, there is a component of the firing rates at the input and output of any motor control system that is continuous and that can be processed by linear filters.

Low-pass filters are filters that selectively transmit low frequencies. The phase lag[2] due to low-pass filtering causes the frequency-dependent delay between the input and output. Fujita modeled the cerebellar cortex

---

[2]The phase angle is given in degrees between 0 and 360 and depends on the signal frequency. A phase lag of 180° would mean that the input and output vectors have opposite directions. The VOR has a phase lag of roughly 180° for a very wide range of frequencies, plus a few degrees that are due to transmission delays and muscle contraction times.

as a low-pass filter. Fujita's idea was that filtering is made possible by the glomerulus, with its Golgi-cell feedback (Fujita 1982). Such low-pass filtering could be a feature of an internal model.

Dynamic internal models, as defined in Chapter 7, simulate the behavior of the "plant," that is, the body part to be moved. An internal model can be either forward or inverse. Forward models predict what will happen to the motor signal when it reaches the plant; inverse models do the inverse of that computation. For timing movements, the cerebellar cortex may use dynamic models that contain linear filters. For example, the cerebellum could implement an inverse model of the oculomotor plant, so that the output of the cerebellar cortex provides a signal that is appropriate to move the eyes (Kawato & Gomi 1992). Here is the key point: in this formulation, the inverse model takes care of timing. This means that adjustments in the inverse model could potentially modify timing.

One possibility is that a module of the cerebellar cortex (such as a microzone) can contain a forward model of an effector, such as an arm, together with a given situation or context (say, a hammer and a nail) (Wolpert *et al.* 1998). In each of these motor modules, a forward model would be paired with the appropriate inverse model. The job of the inverse model is to generate the appropriate motor command, including timing, while the job of the forward model is simply to predict the behavior of the system.

A microzone is the set of P-cells, with their associated circuitry, that receive identical sensory information from the inferior olive. This means that the microzone performs a transformation on a nicely circumscribed set of data. If the microzones correspond to internal models, then inverse dynamic models may exist, in different cerebellar microzones, for the limbs, fingers, head, eyes, and so on. To generate timed movements, the inverse models are trained during experience by motor feedback errors that are conveyed by the firing probability of climbing fibers.[3] When errors occur, they indicate that the movement is inappropriately timed or the wrong amplitude, and the inverse model should be modified by learning. This would be an adaptive filter.

## Multiple channels

In order to endow a motor system with flexibility in different situations (for smooth movements), different versions of each adaptive filter, with different dynamic properties, are needed (Lisberger *et al.* 1983). These could be arrayed in the cerebellar cortex, taking advantage of its processing power. Signals could diverge into the different channels of

[3] Although the firing rate itself is too slow to encode motor error, the probability of firing is a continuous function of time and can encode motor error.

this filter and be weighted separately according to experience, then recombined to control a common motor system, in a form of expansion recoding. This arrangement would require separate channels of processing in the cerebellar cortex on a small scale. The multiple channels of the adaptive filter for a given body part could correspond to units within microzones whose synaptic "weights" can be independently adjusted (Dean *et al.* 2010). The output of the array of channels could be combined to generate a very large number of combinations, for example, for accelerating sequences of movements.

How prevalent are the kinds of signals that can be acted upon by linear filters? In fact, they are very common. Continuous signals are necessary for all motor systems, not only because we need muscle tone to remain standing but also because we often interact with moving objects in the outside world, whether we are tracking the object or trying to capture it. Early studies showed that stimulation of the parallel fibers caused P-cells to increase their discharge rates in a linear fashion, so that sensory and motor information can be relayed without distortion (Eccles *et al.* 1967). But discrete responses are also necessary for most motor systems. Pure delays can be learned not only by the NMR, but also by saccades, and even the smooth pursuit system (Medina *et al.* 2005).

The vestibulo-ocular reflex (VOR), on the other hand, is a brainstem reflex and, apparently, it is not smart enough to learn pure delays; it seems to be constrained by the rules of linear filters. It can learn a new phase lag. For example, the visual field and the head might move sinusoidally 90° out of phase with each other, and the VOR will learn a phase lag or lead to stabilize gaze on the moving field. But unlike timing and amplitude of the NMR, changes in VOR phase are not independent of VOR gain. When we explicitly induce changes in the VOR phase, a learned phase lag causes an increase in VOR gain, while a phase lead causes a decrease in gain. The link between timing and amplitude seems to be obligatory. For example, a mouse mutant which has a deficit in a particular calcium channel, the P/Q channel, shows selective deficits in the learned phase of the VOR. At lower frequencies, where the inverse model is especially important in generating the VOR's response, gain changes are also impaired (Katoh *et al.* 2008). The link between phase and gain may reflect the implementation of an inverse model as a linear filter by the same premotor neurons that regulate gain. Perhaps because the VOR pathway is so simple anatomically, it only relies on linear filters, and it does not have the capability of modifying its timing and amplitude separately.

## *Pattern recognition and motor learning*

Linear models of cerebellar function, as well as the neural-network models we have examined so far, assume that the important signals are neuronal discharge rates, that is, scalar variables that are continuous functions of time. However, changes in the discharge rate may be too gradual to do the job of timing in some cases. Challenging motor tasks (like saccades) require timing on a millisecond time scale. Because cerebellar neurons generally fire at less than 200 Hz, this means that there may be time for only one or two spikes during the saccade. But one spike cannot define a firing rate. That is why experimental measurements of the discharge rate must rely either on smoothing, which generates a continuous function, or averaging over many trials. The (generally unspoken) assumption is that averaging across neurons can take the place of averaging over trials. But what if it does not? What if the system works in a fundamentally different way, such that the exact timing of spikes is important information?

The exact timing of spikes may be important if action potentials are synchronous across neurons. Synchrony can increase the probability of transmission across the next synapse; we will see some examples shortly. But first, let's consider synchrony as a feature of the overall discharge pattern in time and space. Nonlinear networks have been important for simulating discrimination among discharge patterns. Some of these models come a bit closer to the characteristics of actual neurons than the linear models do.

Pattern discrimination was explored in a compartmental model of P-cells in a neural network, by Volker Steuber, Erik De Schutter, and others (Steuber *et al.* 2007). They studied recognition of patterns that depended on cerebellar LTD. These patterns consisted of synchronous activation of a subset of the P-cell's parallel-fiber inputs. Even before any learning had occurred, such patterns resulted in a burst of spikes in the P-cell (due to many excitatory inputs at once), followed by a pause in firing that was caused by calcium-dependent potassium channels. In Steuber's model, the recognition of a pattern that had already been "learned" by the P-cell resulted in shortening of the pause (as well as a reduction in the burst). This could be a timing mechanism in real life.

Spike timing is an important feature of discharge patterns that has been neglected in the linear models and also ignored in most of the literature on the cerebellum. It is possible that spike timing holds the key to improving the precision of movement timing, especially when the movements are so rapid that firing rate becomes a limitation.

A different encoding mechanism was proposed by Joy Walter and Kamran Khodakh (Walter & Khodakhah 2009). Using a neural network similar to Steuber's, but allowing the output of the network to be a linear function of the inputs, Walter and Khodakh found that the change in this linear function due to LTD also could effectively encode learning; it was not necessary to use pauses. In fact, single P-cells were able to discriminate among learned patterns considerably better using linear encoding than using pauses in their simulations.

How many specific, complex patterns can really be learned by a P-cell? The pattern-recognition capacity of a P-cell depends on both the number of inputs that it receives, which is believed to be roughly 100,000 from parallel fibers, and the number of inputs that are needed to constitute a "pattern." Walter and Khodakh estimated that a minimum of 185 inputs might be the minimum for a pattern, which suggested that as many as 75 patterns could be encoded using the same physical network or group of P-cells. How many networks are there? This number is not really known. Parasagittal microzones of cerebellar cortex probably number around 5000, but the actual functional units may number many more. The intuitive objection that the number is finite, no matter how large, while we can learn an infinite number of behaviors, is wrong. We cannot learn an infinite number, just a very large number. And we can combine functional units to generate behaviors. It is not unreasonable that functional units really do encode movement components that have been learned based on pattern recognition.

## Ending the saccade on time

Although position (of the body, eyes, etc.) is a continuous signal, I have already mentioned that many movements in real life are discrete, and stopping them on time is essential. Forward models may be useful for this, nevertheless. When we reach for a target, for example, we can use forward models to generate prompt feedback about the predicted location of the hand and stop the movement when we estimate that we have reached the target.

In the saccade system, which generates discrete movements, timing mechanisms are especially important. This is because saccades must stop before there is time for sensory feedback (visual feedback would require nearly 100 ms, while the average saccade only lasts about 50). Forward models could permit saccades, and also gaze shifts involving the head, to land at least approximately on target. The command signal to the motoneurons during a saccade depends on the motor error, which is the angle between the current gaze direction and the goal. Measuring the actual error would take too long, so the brain uses an estimate of motor error generated by a forward model of the oculomotor plant, a kind of virtual motor error. The forward model may be located in the cerebellar cortex (Quaia *et al.* 1999). When virtual motor error reaches zero, so can the drive to the motoneurons. Modifying the forward model could result in adjustments of saccade amplitude by changing the time at which the drive drops to zero. However, if this were the only mechanism for stopping the saccades, they would overshoot slightly; the muscle is already contracted and pulling, and the eye is moving at a high speed. That is why actually stopping the saccade on target is an active process.

The active process that stops saccades seems to depend on the fastigial nucleus. P-cells on the side of lobule VI or VII that the saccade is directed toward fire a burst of action potentials that reaches its highest discharge rate just before the end of the saccade. During the burst, fastigial neurons on the same side are silenced (Ohtsuka & Noda 1995). When the burst stops, the fastigial neuron exhibits an immediate, intense rebound burst of action potentials. Without the burst in fastigial neurons, the saccade still ends, but it does not end on target (Quinet & Goffart 2005). When a monkey learns to increase saccadic gain, the duration of the saccade increases and more P-cells begin to respond late in the saccade, possibly extending it (Catz *et al.* 2008). Learning a larger saccade size also delays the terminating burst of spikes in fastigial neurons.

By activating a group of "inhibitory burst neurons" in the reticular formation, the fastigial burst of firing inhibits the motoneurons, causing the muscle to relax. Some fastigial neurons activate "excitatory burst neurons" that cause the antagonist muscle to contract. Using this two-pronged approach, the burst from fastigial neurons ensures that the saccade ends on time. A similar mechanism seems to operate even if there is a head movement contributing to the gaze shift (Fuchs *et al.* 2010). In other words, the head is included in a forward model, and is also stopped by the fastigial nucleus. This system has great specificity and flexibility. Whether a particular saccade amplitude is used or not depends on whether the current circumstances match the context in which it was learned (Shelhamer & Clendaniel 2002).

Christian Quaia has proposed a mechanism that could generate the terminating burst by timed spread of excitation across the midline of the vermis during the saccade, by way of the parallel fibers (which can be several mm long) (Quaia *et al.* 1999). In this scheme, P-cells fire their bursts sequentially across the vermis from one side to the other, and when the fastigial neurons on the opposite side receive an adequate input burst, they stop the saccade. Quaia's idea is a variant of the "delay line" idea proposed long ago by Braitenberg. Parallel fibers may be too slow, however. The midline location of the vermis helps, and so does the fact that coordination of the two eyes is accomplished downstream (Chapter 8).

Clearly, there is a lot we can do just with discharge rates. If we use the basic concepts of linear filters and add a few features, such as rebound firing and perhaps delay lines—which can also be linear—we might be able to stop movements such as saccades accurately. Adding synchrony to the discharge pattern across a population of parallel fibers can help with pattern discrimination, motor learning, and precision control of speed and amplitude. But we still need a mechanism of timing across joints, and across limbs. This is absolutely necessary for coordination. For example, let's consider the standard sobriety test, tandem walking. Tandem, or heel-to-toe, gait is what we use when we walk along a line or on a balance beam. A parallel-fiber mechanism has been proposed for timing movements of the two feet for tandem gait. Amy Bastian has found that cutting parallel fibers that normally cross the midline interferes with the mechanism of coordination. The deficit in tandem gait can be obtained without the more conventional vestibular deficits, if the lesion is only of the midline vermis (Bastian *et al.* 1998). It was later proposed that tandem gait is coordinated by synchronous activation of P-cells by the same parallel fibers (Thach 2007). But as with saccades, a problem remains: Realistically, parallel fibers probably can't time anything. They are unmyelinated, and there are hundreds of thousands of them converging on each P-cell. We need to think of something else.

## TIMING USING SYNCHRONOUS FIRING

The pattern of activity in the cerebellum has two more features that we have ignored so far. They may be of key importance in motor timing and coordination. These features are synchrony among the climbing fibers, and synchrony of the simple spikes that convey information out of the cerebellar cortex.

### *The climbing-fiber clock*

As Braitenberg originally suggested (Chapter 2), climbing fibers may play a key role in cerebellar timing. Using the timing of synchronous spikes within a group of climbing fibers as a reference clock would make the best use of neurons in the inferior olive, which fire at low rates and are connected to each other by gap junctions. In general, synchrony does not require cells to fire reliably, only to fire at the correct time when they do fire. Even if not all of the cells are firing at once, enough of them may fire at once to have important downstream effects.

As we would expect from synchrony in climbing fibers, complex spikes are often synchronous within groups of P-cells (Llinas & Sasaki 1989; Lang *et al.* 1999). Electrotonic coupling between groups of neurons in the inferior olive means that they effectively fire as a unit, which is probably the main basis for the climbing-fiber synchrony. But at the level of the P-cell, the degree of synchrony in complex spikes is modulated by sensory stimulation and is not obligatory, suggesting that other inputs to the P-cells (such as parallel fibers and basket cells) may affect the synchrony of complex spikes (Wise *et al.* 2010).

Complex-spike synchrony generates a rhythm that superficially resembles a clock, ticking at roughly 7 times per second (Welsh *et al.* 1995). However, rather than measuring intervals, the clock ticks are thought to generate timed samples. The P-cells that fire together, and are in the same sample group, tend to be part of a functional module and are located in a narrow parasagittal band. Each band is less than 100 $\mu$m across; there are usually about six P-cells in a band. Several could fit in a microzone, as microzones are usually defined (Ozden *et al.* 2009). Each synchronous module includes coupled inferior olivary neurons, the P-cells that are associated with them, their associated basket cells and Golgi cells, and inhibitory neurons in the DCN that receive input from the P-cells and connect back to the inferior olive (De Zeeuw *et al.* 2011). The basket cells, whose axons extend parasagittally, may actually help to maintain complex-spike synchrony within a module. The inhibitory input to the inferior olive from the deep nuclei may also regulate coupling via gap junctions. Each of these components seems to contribute to keeping the clock ticking.

## *The functional significance of synchrony*

Complex spikes show greater synchrony in the presence of sensory stimuli or motor activity that is significant to the particular module. Although individual inferior olivary neurons do not respond reliably to sensory events, sensory events do cause synchronous activation of complex spikes, perhaps, by means of the other inputs to P-cells. The synchronous activity codes the sensory event with high reliability (Ozden *et al.* 2009).

Synchrony affects movements by way of the DCN. Complex spikes are transmitted as short bursts of spikes to the DCN, and inhibit nuclear neurons. The convergence of hundreds of synchronous bursts on single DCN neurons would be expected to silence them briefly. The pause in simple spikes, which follows the complex spike in each P-cell, would then allow strong rebound excitation in the DCN neurons. In Chapter 4, I argued that this sequence of events would cause learning in the deep nuclei. Possibly, synchronous complex spikes can cause learning in the deep nuclei at the same time as LTD in the cerebellar cortex, using the same sensory signal.

For ongoing behavior, the important fact may be that the first spike after the pause caused by inhibition is always very precisely timed (Gauck & Jaeger 2000). As already mentioned, this timing results in stopping movements accurately. Some DCN neurons connect to premotor areas (such as the thalamus and red nucleus), and others (like vestibular neurons) project directly to motoneurons. This suggests a causal relationship, by way of the DCN, between complex spikes and the timing of ongoing movements and of stopping movements like saccades. Gap junctions and olivary neuronal synchrony are necessary for expressing the learned timing of eye-blink reflexes, as well as for correct timing of locomotor movements on the rotorod (van der Giessen *et al.* 2008), extending the importance of complex-spike synchrony to a wide range of movement types.

In general, synchronous complex spikes may contribute to the timing of movement components. Llinas and coworkers showed that the synchronous activation of climbing fibers can represent different components of a licking movement in different populations of P-cells (Welsh *et al.* 1995). Building on this idea, de Zeeuw and colleagues argue that the excitatory and inhibitory inputs to the inferior olive control different sets of microzones that contribute different components of a movement.

It is especially interesting that olivary neurons can remain synchronous during the execution of the same movement at different speeds, suggesting that they are consistently linked to components of a particular task (De Zeeuw *et al.* 2011). This observation suggests a completely different basis for functional units than the array of linear filters that we explored earlier. Linear filters are selective for particular speeds of task execution. How do we reconcile these two principles? It is possible that modules associated with a particular movement component contain sub-modules within the cortex that perform filtering. All of the linear filters associated with a particular movement would then be linked by their climbing-fiber inputs, and selectivity among the filters could be based on sensory input via mossy fibers (for example). That said, linear filters may not be helpful for timing of discrete movements. This means that the nonlinear mechanisms of complex-spike synchrony and of simple-spike synchrony may be more important in many situations.

A cerebellar clock that makes use of rebound firing by the DCN as well as complex-spike synchrony was proposed by Werner Kistler and Chris de Zeeuw in 2002 (Kistler & De Zeeuw 2002). The idea is that post-inhibitory rebound in deep cerebellar neurons can cause reverberating activity in the closed loop encompassing the inferior olive, cerebellar cortex, and deep nuclei. In this class of models, signals are encoded by the exact timing of individual spikes. After a complex spike, the P-cell pauses its firing, and the deep nuclear neuron receiving input from that P-cell generates post-inhibitory rebound and a precisely timed spike that is transmitted to the inferior olive. The timed activity can then reverberate around the loop, taking advantage of the gap junctions in the inferior olive to synchronize the reverberations. The reverberations have a period of roughly 100 ms, which is slow. However, Kistler and de Zeeuw showed that this reverberating activity can operate in different modes. The variety of possible patterns of activity means that the reverberating network can encode more information. It could serve as a basis for working memory in the cerebellum, because activity would persevere within the loop (Kistler & De Zeeuw 2002).

## The cerebellar cortical clock

P-cells fire simple spikes at high rates that could encode considerable information. Neuronal discharge patterns, including simple-spike patterns, are what is filtered by the linear filters. How do these patterns fit with the idea of reverberating loops? It turns out that the pattern of simple-spike activity can interact with the cerebellar loop to time discrete movements (like the NMR) accurately. For example, in a "lock and key" model developed by Mark Schnitzer and colleagues (Wetmore *et al.* 2008), the dynamics of rebound in the deep nuclei could affect the recall of memory by DCN signaling. DCN neurons will respond only if they receive a signal from P-cells that coincides with their rebound activity. The rebound mechanism serves to attenuate poorly timed learned motor responses and amplify correctly timed ones.

Like the learned motor responses, a sensory stimulus needs to have the correct dynamics in order to generate a significant response in the DCN neuron. If rhythmic stimuli of the correct frequency are used, for example, one spike will be generated per cycle with high reliability, precisely encoding the time of each cycle with a small number of actual spikes. This effect is known as "phase locking." Phase locking has been reported in the DCN (Person & Raman 2012) and in neurons of the vestibular nuclei, during the VOR (Broussard *et al.* 2004). Although some stimuli may not be rhythmic in everyday life, the frequencies that they contain can still be encoded in a similar fashion to rhythmic stimuli. Of course, timed movements are easier when they are generated in phase with rhythmic sensory stimuli (as when dancing to music, for example), and we can guess that phase locking in the DCN might contribute to making them easier.

The observation that single-pulse stimuli to the cerebellar cortex can influence spike timing in the deep nuclei (Hoebeek *et al.* 2010) suggests that simple spikes can be significant in controlling the timing of deep nuclear spikes. In everyday life, though, there are no stimulating electrodes to activate all of the P-cells at once; this is easy to achieve in a slice preparation, but are simple spikes ever synchronous in real-life situations? One might think not; simple spikes are thought to reflect the integration of the synaptic inputs to the P-cell with its intrinsic activity, and it is difficult to imagine synchrony in this mixture of inputs. Simple-spike firing in the P-cells of alert animals seems irregular (i.e., it has highly variable interspike intervals). However, close examination of the spike trains reveals that they do fire regularly; this can be obscured by frequent changes in the firing rate (De Zeeuw *et al.* 2011). Even more interesting, these regular patterns are more common during sensory stimulation to which the P-cell is sensitive (Shin *et al.* 2007).

Another surprise has been that the inputs to P-cells are precisely timed. Feedforward inhibition via Golgi cells can limit the input to a given P-cell from its parallel fibers to a brief time window. Synapses between mossy fibers and granule cells in the granular layer have been proposed to transmit spike timing with high accuracy, controlling reliably whether or not a parallel-fiber spike will fall within the window (D'Angelo & De Zeeuw 2009). Feedforward inhibition further sharpens the timing of evoked responses to parallel-fiber stimulation. The feedforward inhibition is necessary for learned changes in VOR phase as well as for consolidation of changes in gain (Wulff *et al.* 2009). Its importance for consolidation may reflect an effect of Wolpert's lock-and-key mechanism, and once again, synchrony across P-cells would be required.

Like complex spikes, simple spikes show a high degree of synchrony. Among neighboring P-cells, 85% fire synchronously without synchronous stimulation; that is, they fire within a millisecond or less of each other. The synchrony is completely unrelated to complex spikes. It seems to be due, in large part, to recurrent inhibition by P-cells, targeting neighboring P-cells (de Solages *et al.* 2008). Thus by an extremely simple and elegant mechanism, neighboring P-cells can fire synchronously. Other mechanisms besides recurrent collaterals also contribute. There are gap junctions among inhibitory interneurons in the molecular layer, as well as between interneurons and P-cells. Cholinergic inputs are also involved in generating the rhythm in some regions, such as lobules VI and VII hemispheres (crus I and crus II) (Middleton *et al.* 2008).

Even though granule-cell inputs are precisely timed, synchronous firing of simple spikes is also a feature that arises in the cerebellar cortex, independent of parallel-fiber inputs. P cells that are more than a few hundred micrometers apart do not fire synchronously (de Solages *et al.* 2008), likely, because P-cell recurrent collaterals extend for only a few hundred micrometers. On the other hand, P-cells with synchronous simple spikes also tend to have synchronous complex spikes and, furthermore, pauses in simple-spike activity are correlated in the same pairs of P-cells. These "coupled P-cells" also tended to have similar responses to sensory stimuli (Wise *et al.* 2010).

P-cell axon collaterals are known to remain confined to zebrin compartments, which are thought to be microzones. Synchrony of simple-spike discharges may provide separate channels for signals that have distinct functional significance, such as a learned change in timing of an eyeblink or a component of a learned movement.

Stellate cells fire in synchrony with oscillations at around 100 Hz that are similar to the cerebral cortical "very fast oscillations" (Middleton *et al.* 2008). Because the synchronous oscillations occur at such high frequencies, they can time intervals of less than 10 ms. This suggests that simple-spike rhythm may act as a clock that times intervals, and determines the resolution at which movements can be timed.

## CONCLUSIONS

The cerebellum produces timed outputs, and processes timed inputs in several different ways. It can do both of these things simultaneously. It is possible that some regions of the cerebellum (such as the vestibulocerebellum) may be specialized for a particular mode of processing (such as linear filtering). Linear filters do appear to be used, and a large number of them may be available in the cerebellum. However, when movements or sensory stimuli are discrete, other modes may be more useful. Timing may be accomplished using a reference clock that is based on synchronous climbing-fiber activity and on reverberating activity in the cerebellar loops. Not only the complex spikes, but also the simple spikes of P-cells are often synchronous. Both complex- and simple-spike synchrony could be used to associate information that relates to particular movements.

## REFERENCES

Bastian, A.J., Mink, J.W., Kaufman, B.A. & Thach, T.W. (1998) Posterior vermal split syndrome. *Ann. Neurol.*, **44**, 601–610.

Broussard, D.M., Priesol, A.J. & Tan, Y.-F. (2004) Asymmetric responses to rotation at high frequencies in central vestibular neurons of the alert cat. *Brain Res.*, **1005**, 137–153.

Buonomano, D.V. & Mauk, M.D. (1994) Neural network model of the cerebellum: temporal discrimination and the timing of motor responses. *Neural Comput.*, **6**, 38–55.

Catz, N., Dicke, P.W. & Thier, P. (2008) Cerebellar-dependent motor learning is based on pruning a Purkinje cell population response. *Proc. Natl. Acad. Sci. USA*, **105**, 7309–7314.

D'Angelo, E. & De Zeeuw, C.I. (2009) Timing and plasticity in the cerebellum: focus on the granular layer. *Trends Neurosci.*, **32**, 30–40.

de Solages, C., Szapiro, G., Brunel, N., Hakim, V., Isope, P., Buisseret, P., Rousseau, C., Barbour, B. & Lena, C. (2008) High-frequency organization and synchrony of activity in the Purkinje cell layer of the cerebellum. *Neuron*, **58**, 775–788.

De Zeeuw, C.I., Hoebeek, F.E., Bosman, L.W.J., Schonewille, M., Witter, L. & Koekkoek, S.K.E. (2011) Spatiotemporal firing patterns in the cerebellum. *Nat. Rev. Neurosci.*, **12**, 327–344.

Dean, P., Porrill, J., Ekerot, C.-F. & Jorntell, H. (2010) The cerebellar microcircuit as an adaptive filter: experimental and computational evidence. *Nat. Rev. Neurosci.*, **11**, 30–43.

Eccles, J.C., Ito, M. & Szentagothai, J. (1967) *The Cerebellum as a Neuronal Machine*. Springer, Berlin, Germany.

Fuchs, A.F., Brettler, S. & Ling, L. (2010) Head-free gaze shifts provide further insights into the role of the medial cerebellum in the control of primate saccadic eye movements. *J. Neurophysiol.*, **103**, 2158–2173.

Fujita, M. (1982) Simulation of adaptive modification of the vestibulo-ocular reflex with an adaptive filter model of the cerebellum. *Biol. Cybern.*, **45**, 207–214.

Gauck, V. & Jaeger, D. (2000) The control of rate and timing of spikes in the deep cerebellar nuclei by inhibition. *J. Neurosci.*, **20**, 3006–3016.

Hesslow, G. & Ivarsson, M. (1994) Suppression of cerebellar Purkinje cells during conditioned responses in ferrets. *Neuroreport*, **5**, 649–652.

Hoebeek, F.E., Witter, L., Ruigrok, R.J. & De Zeeuw, C.I. (2010) Differential olivo-cerebellar control of rebound activity in the cerebellar nuclei. *Proc. Natl. Acad. Sci. USA*, **107**, 8410–8415.

Holmes, G. (1917) The symptoms of acute cerebellar injuries due to gunshot injuries. *Brain*, **40**, 461–535.

Ivry, R.B. & Keele, S.W. (1989) Timing functions of the cerebellum. *J. Cogn. Neurosci.*, **1**, 136–152.

Katoh, A., Chapman, P.J. & Raymond, J.L. (2008) Disruption of learned timing in P/Q calcium channel mutants. *PLoS One*, **3**, e3635.

Kawato, M. & Gomi, H. (1992) The cerebellum and VOR/OKR learning models. *Trends Neurosci.*, **15**, 445–453.

Kistler, W.M. & De Zeeuw, C.I. (2002) Dynamical working memory and timed responses: the role of reverberating loops in the olivo-cerebellar system. *Neural Comput.*, **14**, 2597–2626.

Lang, E.J., Sugihara, L., Welsh, J.P. & Llinas, R. (1999) Patterns of spontaneous Purkinje cell complex spike activity in the awake rat. *J. Neurosci.*, **19**, 2728–2739.

Lisberger, S.G., Miles, F.A. & Optican, L.M. (1983) Frequency-selective adaptation: evidence for channels in the vestibulo-ocular reflex? *J. Neurosci.*, **3**, 1234–1244.

Llinas, R. & Sasaki, M. (1989) The functional organization of the olivocerebellar system as examined by multiple Purkinje cell recordings. *Eur. J. Neurosci.*, **1**, 587–602.

Medina, J.F., Garcia, K.S., Nores, W.L., Taylor, N.M. & Mauk, M.D. (2000) Timing mechanisms in the cerebellum: testing predictions of a large-scale computer simulation. *J. Neurosci.*, **20**, 5516–5525.

Medina, J.F., Carey, J.P. & Lisberger, S.G. (2005) The representation of time for motor learning. *Neuron*, **45**, 157–167.

Middleton, S.J., Racca, C., Cunningham, M.O., Traub, R.D., Monyer, H., Knopfel, T., Schofield, I.S., Jenkins, A. & Whittington, M.A. (2008) High-frequency network oscillations in cerebellar cortex. *Neuron*, **58**, 763–774.

Ohtsuka, K. & Noda, H. (1995) Discharge properties of Purkinje cells in the oculomotor vermis during visually guided saccades in the macaque monkey. *J. Neurophysiol.*, **74**, 1828–1840.

Ozden, I., Sullivan, M.R., Lee, H.M. & Wang, S.S.-H. (2009) Reliable coding emerges from coactivation of climbing fibers in microbands of cerebellar Purkinje neurons. *J. Neurosci.*, **29**, 10463–10473.

Person, A.L. & Raman, I.M. (2012) Purkinje neuron synchrony elicits time-locked spiking in the cerebellar nuclei. *Nature*, **481**, 502–505.

Quaia, C., Lefevre, P. & Optican, L.M. (1999) Model of the control of saccades by superior colliculus and cerebellum. *J. Neurophysiol.*, **82**, 999–1018.

Quinet, J. & Goffart, L. (2005) Saccade dysmetria in head-unrestrained gaze shifts after muscimol inactivation of the caudal fastigial nucleus in the monkey. *J. Neurophysiol.*, **93**, 2343–2349.

Shelhamer, M. & Clendaniel, R.A. (2002) Context-specific adaptation of saccade gain. *Exp. Brain Res.*, **146**, 441–450.

Shin, S.-L., Hoebeek, F.E., Schonewille, M., De Zeeuw, C.I., Aertsen, A. & De Schutter, E. (2007) Regular patterns in cerebellar Purkinje cell simple spike trains. *PLoS ONE*, **2**, e485.

Steuber, V., Mittman, W., Hoebeek, F.E., Silver, R.A., De Zeeuw, C.I., Hausser, M. & De Schutter, E. (2007) Cerebellar LTD and pattern recognition by Purkinje cells. *Neuron*, **54**, 121–136.

Thach, W.T. (2007) On the mechanism of cerebellar contributions to cognition. *Cerebellum*, **6**, 163–167.

Timmann, D., Watts, S. & Hore, J. (1999) Failure of cerebellar patients to time finger opening precisely causes ball high–low inaccuracy in overarm throws. *J. Neurophysiol.*, **82**, 103–114.

van der Giessen, R.X., Koekkoek, S.K., van Dorp, S., De Gruiji, J.R., Cupido, A., Khosrovani, S., Dortland, B., Wellershaus, K., Degen, J., Deuchars, J., Fuchs, E.C., Monyer, H., Willicke, K., De Jeu, M.T.G. & De Zeeuw, C. (2008) Role of olivary electrical coupling in cerebellar motor learning. *Neuron*, **58**, 599–612.

Walter, J.T. & Khodakhah, K. (2009) The advantages of linear information processing for cerebellar computation. *Proc. Natl. Acad. Sci. USA*, **106**, 4471–4476.

Welsh, J.P., Lang, E.J., Sugihara, I. & Llinas, R. (1995) Dynamic organization of motor control within the olivocerebellar system. *Nature*, **374**, 453–457.

Wetmore, D.Z., Mukamei, E.A. & Schnitzer, M.J. (2008) Lock-and-key mechanisms of cerebellar memory recall based on rebound currents. *J. Neurophysiol.*, **100**, 2328–2347.

Wise, A.K., Cerminara, M.L., Marple-Horvat, D.E. & Apps, R. (2010) Mechanisms of synchronous activity in cerebellar Purkinje cells. *J. Physiol.*, **588**, 2373–2390.

Wolpert, D.M., Miall, R.C. & Kawato, M. (1998) Internal models in the cerebellum. *Trends Cogn. Sci.*, **2**, 338–347.

Wulff, P., Schonewille, M., Renzi, M., Viltono, L., Sassoe-Pognetto, M., Badura, A., Gao, Z., Hoebeek, F.E., van Dorp, S., Wisden, W., Farrant, M. & De Zeeuw, C.I. (2009) Synaptic inhibition of Purkinje cells mediates consolidation of vestibulo-cerebellar motor learning. *Nat. Neurosci.*, **12**, 1042–1049.

## FURTHER READING

Ivry, R.B. & Keele, S.W. (1989) Timing functions of the cerebellum. *J. Cogn. Neurosci.*, **1**, 136–152.

Kawato, M. & Gomi, H. (1992) The cerebellum and VOR/OKR learning models. *T. I. N. S.*, **15**, 445–453.

Mauk, M.D. & Buonomano, D.V. (2004) The neural basis of temporal processing. *Ann. Rev. Neurosci.*, **27**, 307–340.

D'Angelo, E. & De Zeeuw, C.I. (2009) Timing and plasticity in the cerebellum: focus on the granular layer. *T.I.N.S.*, **32**, 30–40.

Dean, P., Porrill, J., Ekerot, C.-F. & Jorntell, H. (2010) The cerebellar microcircuit as an adaptive filter: Experimental and computational evidence. *Nat. Rev. Neurosci.*, **11**, 30–43.

# Section IV

## Interpreting the World

Our experience tells us that movement and thought are fundamentally different things. Thought is inner experience; movement is how we interact with the outside world. We can have thought without movement, and vice versa. The cerebellum is a brain structure that originally evolved for motor control, not for thought.

These statements are all undoubtedly correct. But the human cerebellum is also a cognitive structure. At the level of the encoded representation in the brain, both thought and movement are represented using neurons and action potentials. In the brain, there is no fundamental difference between thought and movement (Ito 1993). So why not use the cerebellum to improve our cognition as well as our motor control? Although in the past this idea has not been popular, scientific opinion is shifting. Most cognitive tasks would make good use of a computer, if only we had one that was properly programmed. Our built-in computer, the cerebellum, may perform computations that help us to select the right word for a rhyme while writing a poem, calculate distances during drawing and maze tasks, or classify objects as similar or different. Perhaps the cerebellar computer comes in handy for cognitive tasks that are computation-intensive.

For motor systems, a small army of neuroscientists has worked out, over the decades, many details of how the cerebellum makes its contribution. The situation is very different for non-motor systems, however. The authors whose work I will summarize in this section feel that there is no longer any doubt that it does contribute to cognitive function. On the other hand, these interpretations have been controversial, and have even sustained direct attacks (Glickstein 2006). As noted by the critics, in the field of cerebellar cognition, attempts to replicate results sometimes fail. For one thing, tests for most cognitive functions are well established only for humans; this means that lesions are always due to some pathology and are not reproducible. Also, motor control unavoidably plays a part in every response. Without motor timing and coordination, which we know are cerebellar functions, the most intelligent person in the world would

*The Cerebellum: Learning Movement, Language, and Social Skills*, First Edition. Dianne M. Broussard.

be unable to speak, write, draw, or handle cards, and would fail all of the cognitive tests. Lesser degrees of motor impairment might systematically reduce scores. Proper controls for these effects are often omitted. But in spite of these issues, remarkable progress has been made.

## THE HUMAN CEREBELLUM

An enormous expansion of the cerebellar hemispheres, particularly the posterior lobe, occurred during the evolution of humanoid lineages. The posterior lobe is not important for bipedal locomotion, but some of the expansions likely have to do with coordination of the hands, arms, and fingers in order to manipulate and throw objects. Some non-motor functions have also been strongly implicated.

Along with the posterior lobe, the size of the dentate nucleus expanded several hundredfold during the evolution of the primates, especially the great apes. A new region of the dentate, resembling a folded cortex, appeared. The forebrain also expanded quickly. In earlier members of the genus *Homo*, the cerebral hemispheres were large relative to the cerebellum (larger, relatively, than in other mammals). Even Neanderthals, an early variant of *Homo sapiens*, had a relatively small cerebellum. But finally, the cerebellum did catch up. In modern humans, the cerebellum is relatively significantly larger and appears to have returned to the average ratio of cerebellum to cerebral hemispheres that is seen in other mammals[1] (Weaver 2005).

The latest expansion of the cerebellum at least partly coincided with the appearance of human technology. The Neanderthals had a culture that included personal ornamentation, burial of the dead, and representational art. But they were lacking advanced tools such as bows, weirs, loom weights, eyed needles, and everything else. Susan Weaver has speculated that more advanced technologies may have been linked to greater social complexity, and that the cerebellar expansion was advantageous for both. In modern primates, the cerebellum is closely interconnected with the forebrain. The cerebellum is also more efficient than the forebrain in terms both of information capacity and processing speed. In a way analogous to motor planning, it may provide "rule-based, procedural organization of sequential operations across many cognitive domains in response to cultural pressures" (Weaver 2005).

In this section I will present evidence that the cerebellum contributes to several non-motor cognitive and social functions. These functions can be

[1]There is a lot of variation among mammals. Even among bats, the fruit bats have a relatively larger cerebellum, relative to the cerebral cortex, than the insectivorous bats. For more details see Weaver (2005).

evaluated using a wide variety of neuropsychological tests (Stoodley & Schmahmann 2009). Of course, many cognitive functions are inextricably linked to each other. It may not be possible, to separate working memory from executive function, or executive function from language, or working memory from attention. Some distinctions between these functions are supported by localization of function. But others may simply reflect standard interpretations, and perhaps should be questioned. With these caveats, in Chapter 11, I will describe what the cerebellum does for certain aspects of intelligence and social skills. These include executive function, working memory, problem-solving, and language. In Chapter 12 I will discuss the role of the cerebellum in directed attention, spatial ability, sensory perception, and emotion.

# 11 Intelligence and Language

Does the cerebellum have the necessary anatomical connections to be involved in cognitive function? Virtually the entire neocortex, including prefrontal cortex, projects to the pons, providing an enormous input to the cerebellar cortex. The dentate nucleus projects back to the neocortex by way of the thalamus. The neocortex also projects to the red nucleus, which projects heavily to the inferior olive in humans, contributing to the climbing-fiber pathway (Figure 11.1). But a key question is: Which neocortical regions are connected to the cerebellum?

The connections between the cerebellum and the cerebral cortex are arranged in closed loops, and some of these loops have motor functions. Several regions of the cerebellar hemispheres and vermis connect with primary motor cortex (Brodmann's area 4), the region of the cerebral cortex that sends axons directly to the spinal cord. Hemispheric lobules IV, V, VI, VII and VIII all have areas that are connected with motor cortex. The connections from the cerebellar hemispheres to the motor cortex are by way of the less-folded, dorsal part of the dentate nucleus, which is present in all mammals (Kelly & Strick 2003).

On the other hand, the cerebrocerebellar loops that are outlined in Figure 11.1 have no motor function. For example, in monkeys, cerebellar projections, by way of the dentate nucleus and the thalamus, reach Walker's area 46. The human version of area 46 is called the dorsolateral prefrontal cortex. It is an important region for higher cognitive function, and also functions in suppressing inappropriate emotional responses. Damage to the dorsolateral prefrontal cortex is believed to be responsible for the pattern of deficits exhibited by the famous patient Phineas Gage.[2] The region of the cerebellar cortex that is connected with area 46 is crus II, which is greatly expanded in humans (Kelly & Strick 2003).

Crus II is not interconnected with the motor areas of the neocortex. Given the proposed function of crus II in motor planning, which I

[2]Gage survived the passage of a tamping iron through his left frontal lobe after an explosion. Although he was physically able to work after his recovery, Gage could not keep a job due to a profound loss of social skills. His intelligence was also affected.

*The Cerebellum: Learning Movement, Language, and Social Skills*, First Edition. Dianne M. Broussard.

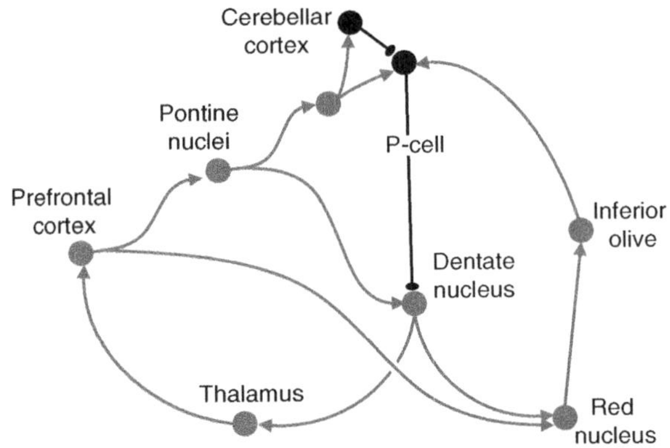


**Figure 11.1** The connections of the human cerebellum that are likely to participate in cognition and language. The prefrontal cortex, including Broca's area, projects both to the pontine nuclei and to the red nucleus. The hemispheres of the cerebellar posterior lobe project to the dentate nucleus and back to the prefrontal cortex by way of the thalamus. The red nucleus provides input to the cerebellum by way of the inferior olive and the climbing fibers.

discussed in Chapter 8, it is surprising that in monkeys, neurons in crus II do not respond during reaching movements, while neurons in the surrounding areas do respond (Middleton & Strick 1994). Perhaps in primates, the newer parts of crus II (i.e., most of crus II) are involved in higher-level planning.

The cerebellum also projects indirectly to the posterior parietal cortex, a cortical region that is activated during nearly all cognitive tasks. The projections to both the parietal and the prefrontal cerebral cortex arise from specialized regions of the ventral dentate nucleus and the interpositus. In primates, including humans, the ventral dentate nucleus is folded and different in appearance from the dorsal dentate nucleus, and is much larger than the dorsal dentate nucleus. The ventral dentate has two distinct regions that connect with parietal and prefrontal cerebral cortex, respectively (Dum & Strick 2003).

The nucleus interpositus, like the dentate, has a specialized ventral region in its posterior subdivision that receives input from the paravermal posterior lobe and projects, via the thalamus, to area 46. However, in the case of the interpositus, the same area that projects to area 46 also projects to premotor regions of the frontal lobe, and to the parietal lobe (Lu *et al.* 2012).

The major language area in the cerebral cortex is Broca's area in the frontal lobe,[3] and Broca's area may be part of another cerebrocerebellar

[3]Broca's area corresponds to Brodmann's areas 44 and 45, which are located in the inferior lateral part of the left or less commonly, the right frontal lobe. Broca's area is thought to be involved in language.

loop. Input to the inferior olive, possibly providing error signals for the cognitive and language systems, arises from the red nucleus of the brainstem. In mammals, the red nucleus is a major premotor nucleus, linking directly to motoneurons in the eyeblink system and limb motor systems. However, in humans the projection from the red nucleus to the inferior olive is larger than in other mammals, and the projection from the red nucleus to the spinal cord is relatively smaller. In humans, the red nucleus receives the bulk of its input from the cerebral cortex, including dorsolateral prefrontal cortex and Broca's area (Leiner *et al.* 1993).

Together, these anatomical findings suggest that different regions of the cerebellar hemisphere and dentate nuclei are processing information for different tasks, and that these include a variety of motor tasks as well as cognition, language and emotional control. We can think in terms of using the computational power of the cerebellum to assist a variety of cortical regions that need help with their tasks. Without the cerebellum, the tasks can still be done, just not as well.

If the non-motor regions of the dentate nucleus and cerebellar hemispheres are involved in cognition, then we should be able to see changes in activity associated with specific cognitive tasks in human subjects. Damage to these regions in humans also should cause cognitive impairment. Last but not least, these effects should not be due to motor aspects of the task.

## GENERAL INTELLIGENCE

Cerebellar damage affects scores on intelligence tests in humans. In fact, the behavioral abnormalities sometimes fit a pattern, known as "cerebellar cognitive affective syndrome." The syndrome was originally documented in 20 patients with lesions that included the cerebellum but not the cerebral cortex. Most of them had strokes affecting the posterior inferior cerebellar artery (PICA) on one or both sides.

The pattern in cerebellar cognitive affective syndrome is compelling. It includes deficits in executive function, attention, abstract reasoning, linguistic ability, and social skills. Executive function in particular is severely affected. Scores on the standard intelligence tests are significantly lower than normal. All of the patients with this syndrome have damage to the posterior lobes (Schmahmann & Sherman 1998). The different characteristics of the syndrome may be due to damage to different parts of the posterior lobe, and the appearance of the entire syndrome may require damage to both the hemispheres and the vermis.

Developmental cerebellar disorders also can affect intelligence. For example, congenital malformations affecting the cerebellar posterior lobe cause mental retardation. Retardation is especially profound if the vermis is mostly missing (vermal agenesis) (Tavano *et al.* 2007). Also, the cerebellum is vulnerable to stress at the time of birth, because it is still at an early stage of development; neuronal migration in the cerebellum is not even complete at birth. Infants that are born prematurely can grow up with low IQ scores and poor working memory, and also can have small cerebella. The impact of premature birth on intelligence could be caused by its impact on cerebellar development (Allin *et al.* 2001). Degenerative disorders, including spinocerebellar degeneration and other types of cerebellar atrophy, can also cause deficits in general intelligence and in cognitive planning (Leiner *et al.* 1993).

In older children, surgery to remove tumors of the posterior fossa[4] can lead to loss of intelligence as measured by standard IQ tests (like Raven's progressive matrices, which is thought to measure the overall intellectual level without requiring good verbal skills, and the Wechsler Adult Intelligence Scale). In a few such studies, IQ has been measured in the same patients before and after the surgery, to confirm a decrease (Cantelmi *et al.* 2008). Relatively small cerebellar lesions cause relatively specific deficits. For example, verbal intelligence is thought to be affected by damage to the right posterior lobe of the cerebellum, while spatial ability may depend on the left hemisphere of the posterior lobe. One study on children, in which motor disabilities were tested and taken into account, found that the posterior-lobe vermis contributes to all aspects of intelligence, including vocabulary, complex language, working memory, and executive function (Steinlin *et al.* 2003).

### *Possible contributing factors*

Motor function can be a factor in low IQ scores, and in fact this possibility is very difficult to rule out. On the other hand, a recent retrospective analysis of 156 adults with either cerebellar atrophy or restricted (stroke or surgical) lesions showed no correlation between the motor and cognitive scores (Tedesco *et al.* 2011). This suggests that the striking cognitive impairments that were found in children with posterior-lobe damage are not due to motor impairments.

Among another group of adults with restricted cerebellar lesions that was studied in Italy, those with posterior-lobe damage due to occlusion of the PICA had significantly lower than normal scores on tests of verbal memory, language, and executive function. Interestingly, the PICA group had a normal mean score in Raven's progressive matrices, suggesting that the problem is less severe for non-verbal measures of intelligence. An analysis of the damaged regions indicated that the hemispheres of lobes VI–IX all contributed to the verbal deficits, with crus II perhaps making

---

[4]The posterior fossa is the compartment that houses the cerebellum and brainstem.

the most consistent contribution (Tedesco *et al.* 2011). In fact, several studies do find that the cerebellum participates in some non-verbal aspects of intelligence, and have tried to tease these factors apart.

One group with well-established cognitive disability consists of patients with autism spectrum disorders. Autism is linked to impairments in language, social skills, cognitive planning and problem solving, as well as other aspects of intellectual function. Difficulties with communication are an important diagnostic criterion for autism, and such difficulties may affect most measures of intelligence.

Cerebellar dysfunction can even contribute to autism. The vermis tends to be small in autistic children, and the Purkinje cell (P-cell) layer is reduced in size, suggesting a reduced number of P-cells, in lobules VI–VII of the vermis. The entire vermis is smaller than normal. Strangely, the abnormally small size of the vermis is correlated with an abnormally *large* size of the frontal lobes of the cerebral cortex (Carper & Courchesne 2000). One way to interpret this is that even with the increased size of the cortical hemispheres, the loss of cerebellar P-cells takes its toll on the ability to communicate.

In fragile X syndrome, which sometimes leads to autism, the posterior vermis (lobules VI–X) is small, both in absolute and in relative terms. Thirty percent of fragile X patients meet the full criteria for autism, while 90% show at least some autistic features (Hampson *et al.* 2012). In fragile X, the reduction in vermal size is correlated with the decreased performance on IQ tests (Mostofsky *et al.* 1998).

In sum, there is a large amount of circumstantial evidence that the posterior lobe of the cerebellum is involved in intellectual function, especially in verbal skills. In the rest of this chapter and Chapter 12, I will consider language skills and several other cognitive functions, and will ask two questions: first, is the cerebellum involved? Second, is there a particular region of the cerebellum that is involved? Positive answers to both questions would support the argument that these aspects of intelligence are mediated by networks that include interconnections between specific cerebellar cortical regions and specific regions of the cerebral cortex. The first aspect of intelligence to be considered is executive function.

## EXECUTIVE FUNCTION

The ability to carry out the cognitive tasks that are necessary to meet a goal, while suppressing behavior that detracts from the goal, is known as executive function. It encompasses multiple functions that contribute to goal-directed behavior, such as planning, working memory, set shifting,[5]

[5]Set shifting is the ability to shift from one task, or set of rules, to another.

and abstract reasoning. It includes global, long-range planning and social skills. Executive function is especially important in dealing with novel, difficult or dangerous situations, or situations where habits need to be overcome or distractions need to be ignored. It is an important component of intelligence.

There is no single cerebral cortical region that is responsible for executive function. The dorsolateral prefrontal cortex (or in monkeys, area 46) is thought to participate. Monkeys actually have a relatively small anatomical linkage between the cerebellum and the prefrontal cortex, but in humans, there is a much larger projection from the (also much larger) prefrontal cortex to the cerebellum (Rammani 2006). Also in humans, the cerebellum is part of the "executive control network," which is thought to retain information that we need for action preparation. The executive network includes the dorsolateral prefrontal cortex and other regions of the prefrontal and parietal cortices. Large regions of the right and left cerebellar hemispheres are part of this network (Habas *et al.* 2009). Across many studies of this network, crus I is the cerebellar region that is most commonly identified as contributing to it (Stoodley & Schmahmann 2009).

Although the cerebellum is certainly part of the executive network, the function of the cerebellum in the network is less clear. The linkages that make up the network can be detected in the resting state; neither a task nor movement is required (Habas *et al.* 2009). Furthermore, tests of decision-making, planning, and other aspects of what is thought to be executive function do not consistently activate crus I or crus II (Stoodley & Schmahmann 2009). On the other hand, cerebellar damage does affect executive function. In patients with cerebellar atrophy, strategic planning for carrying out actions can be abnormally slow. Strategic planning and executive function are also disrupted by surgery for posterior-fossa tumors in children (Steinlin *et al.* 2003). Planning for tasks like drawing complex designs can be severely impaired, and speech can also be slow in such children (Levisohn *et al.* 2000). It is possible that the cerebellum participates in particular aspects of planning, perhaps when a task is particularly challenging. Some tasks may not require it.

As I discussed in Chapter 8, motor planning is an important function of the cerebellum. Although strategic planning and motor planning are quite different from each other, the cerebellum may be involved in both. It may be involved in organizing tasks that range from purely motor (like locomotion, grasping, or pronouncing a word correctly) to tasks that have cognitive components (like making a drawing or finding one's way through a maze) (Schmahmann & Sherman 1998).

## Working memory

Working memory is an important part of executive function, and working-memory tasks are often used in intelligence testing. Working memory allows us to remember what we are doing. It is necessary for performing previously-learned sequences of actions, and for learning new ones (procedural or sequence learning). It is also necessary for reasoning and for language comprehension. Working memory depends on attention and contributes to several cognitive skills related to attention. It is necessary for focusing attention over time as well as for multitasking.

Physically, working memory is reverberating and/or synchronous activity in large cerebral cortical networks that encompass association areas. The lateral prefrontal and posterior parietal regions are the most commonly implicated in working memory. These areas have extensive connections with the cerebellum via the pons and thalamus in humans and when they become active, so does the cerebellum (Habas *et al.* 2009).

There is strong evidence that the cerebellum contributes to working memory. During working memory, information may be stored phonologically,[6] as when we repeat (silently) a telephone number that we are planning to dial. Phonological storage is thought to require a loop of reverberating activity that includes the hemispheric portions of cerebellar lobule VI VIII, including crus I and crus II, and the cerebral cortex (Figure 11.1) (Desmond *et al.* 1997). The regions of hemispheric crus I and crus II that are activated during working memory are adjacent to, but do not overlap with, the regions that are involved in a purely motor finger-tapping task (Stoodley & Schmahmann 2009).

Although working memory does not require the fingers, it does seem to involve some of the motor regions that normally produce speech. This is potentially confusing, because motor planning, which most would not consider a "cognitive" function, might activate areas of the cerebellum during working memory. Is the contribution of the cerebellum to working memory purely motor, then? In one classic working-memory paradigm, subjects are asked to remember a series of letters or numbers for a short period of time. This task activates several areas of the posterior-lobe hemispheres, including crus I and II, more intensely than a control task (silent reading of the same number of letters, followed by an identical motor response). Activation of lobule VII hemispheres in the working-memory task depends on the number of letters to be remembered, while activation in the silent-reading task does not. This means that the activation of lobule VII is related to working memory rather than to the motor aspects of the task (Chen & Desmond 2005).

It is possible that in addition to working memory, the task of reciting remembered letters might involve more motor planning than silent reading does, and some of the extra activation in this task could reflect motor

[6]Phonologically means "by the sound." Inner speech, which involves both motor planning and auditory responses even though we do not actually make a sound, is used for phonological storage.

planning. For example, crus I seems to be involved both in verbal working memory and in phonological rehearsal that does not require memory, suggesting that motor planning could be a factor for the involvement of crus I. On the other hand, crus I is co-activated with Broca's area in the frontal cortex, which is a general language area. Therefore, crus I could be involved in both motor and non-motor aspects of language. In contrast to crus I, parts of lobule VII are co-activated with parietal cortex *only* during verbal working memory, and not during rehearsal, suggesting that motor planning does not require lobule VII (Chen & Desmond 2005). Once again, we can conclude that lobule VII participates specifically in working memory.

Studies on stroke patients also support the role of the cerebellum in working memory. Working memory was affected measurably in two-thirds of the patients that were used to define cerebellar cognitive affective syndrome (Schmahmann & Sherman 1998) and in the large Italian group of stroke patients as well (Tedesco *et al.* 2011). A third study found that patients with PICA lesions perform within the normal range on standard intelligence tests, but poorly on tests that are more selective for working memory, such as the Wechsler Memory Scale—Revised, the trail making test (connecting dots), and others. Episodic and verbal memory are also affected. As already mentioned, PICA lesions affect the posterior lobe of the cerebellum, and the anterior cerebellar lobe appears not to have any of these functions (Exner *et al.* 2004).

## Remembering the order of events

Children who have had surgery to remove tumors from the posterior lobe often perform poorly on the digit-span test, which relies on working memory (Levisohn *et al.* 2000). The digit-span requires subjects to remember a list of numbers in the correct order. In another study, children with damage to the posterior lobe of the cerebellum had difficulty putting events in order in a narrative (Leggio *et al.* 2008). One patient who was diagnosed with cerebellar cognitive affective syndrome had previously been an avid reader of classic literature. After her stroke, she had difficulty reading[7] (Schmahmann & Sherman 1998). Like the digit-span test, literature and storytelling require us to remember the order of events. This ability is known as cognitive sequencing, and it also allows us to keep track of the order of sensory stimuli, motor patterns, or abstract concepts like numbers.

The prefrontal cortex is involved in cognitive sequencing. The cerebellum is clearly involved in motor sequencing (Chapter 7), and may also

[7]Unfortunately it is not clear exactly why this patient had difficulty reading. Her deficit was clearly cognitive and not motor. She may have had trouble following a narrative, which is why I mention her here.

participate in its cognitive counterpart. Both in cognitive sequencing and in predictive motor control, a pattern is created. A narrative pattern consists of events (real or imagined) in the outside world. Although we think of narration as a cortical function, getting the cerebellum involved might improve the speed of recall.

In summary, working memory is necessary for focusing attention over time, for multitasking, for set-shifting (switching from one task to another), and for cognitive sequencing. The cerebellum is involved in at least some of these cognitive skills.

## PROBLEM SOLVING

Since we are proposing that the cerebellum contributes to general intelligence, we should ask whether it is involved in cognitive tasks with a clear goal, such as puzzles, math problems, or mazes. Since problem-solving requires both working memory and executive function, it seems likely that the cerebellum is important for problem-solving. The ability to solve problems is thought to be an important aspect of intelligence.

In a water-maze task that is cued by discriminating between visual patterns (the platforms themselves are invisible), lurcher mice cannot learn to swim to the platform (Lalonde *et al.* 1993). Lesions of the dentate nucleus lead to a similar deficit (Joyal *et al.* 2001). Visuomotor impairment is not a contributing factor in these deficits, because the mice can swim to a visible platform. Other cerebellar mutants with P-cell loss also have poor performance in the water maze with a hidden platform.

Our so-called "sense of direction" may be dependent in part on the posterior vermis, which processes directional motion signals. However, at least some of the deficits in maze performance have been replicated with cerebellar lesions that spared the vermis. The poor performance of the mutant and lesioned mice in the water-maze task could reflect poor spatial learning, attention, or working memory. In left–right alternation tasks, where the mice must remember what they did on the previous trial, performance is more strongly affected than other, similar tasks. This points to a deficit in working memory that could also cause the difficulty in solving mazes (Lalonde & Strazielle 2003).

In humans, cerebellar activity increases during a variety of tests for cognitive function that involve problem-solving, as well as motor planning and other aspects of executive function (Cantelmi *et al.* 2008). In such studies it is always important to control for the motor component of the task. For example, the cerebellum is activated during the Tower of London task. In the Tower of London, subjects are asked to rearrange a pattern of colored balls to match another pattern, by moving the balls individually. The patterns can vary. The part played by the cerebellum in the Tower of London may be purely visuomotor, and not cognitive, since the degree of cerebellar activation does not depend on the complexity of the task (Dagher *et al.* 1999).

Focal cerebellar damage can also have a profound effect on both abstract reasoning and calculation, two important aspects of problem-solving. Of the group of 20 patients that were initially used to define cerebellar cognitive affective syndrome, 14 were deficient in mental arithmetic. One patient in the group had the posterior vermis removed due to a tumor. She had a profound deficit in mathematical ability 2 weeks after her surgery, although she had been a university student before the tumor was found. This coincided with an IQ in the mentally deficient range and performance on a maze task that was at the level of an 8-year-old (she was 22).[8] The patient improved but did not become normal (Schmahmann & Sherman 1998).

Although there are few investigations of mathematics and the cerebellum, the game of chess, which requires logic as well as intuition, tends to increase the activation of the left cerebellar hemisphere (Ito 2008). There is also some evidence that the dentate nucleus is involved in puzzle-solving. If subjects are asked to solve a pegboard puzzle, the dentate nucleus is more active while solving the puzzle, compared with a control task in which the pegs are simply moved from one end of the board to the other. In the control task, which consists of visually guided hand movements, some activation is seen in the dentate on the side corresponding to the hand that is being used. But activity is more intense, more widely distributed within the dentate nuclei, and bilateral, during the puzzle task. This suggests that increased activity in the dentate nucleus is associated with abstract thought (Kim *et al.* 1994).

The ability to think fast is something that we associate with intelligence. In fact, slow processing can cause apparent deficits in problem solving when a time limit is involved, as it often is. Speed is not usually measured apart from the ability to solve the problem. However, reaction time in a memory scanning test is affected by cerebellar damage in multiple sclerosis (Archibald *et al.* 2004). Measurements of speed in problem-solving tasks might reveal a contribution of the cerebellum to thinking fast.

In solving a complicated problem, we often use intuition rather than logic, and the main advantage of this strategy is speed. Recently, Masao Ito has proposed that the cerebellum contributes significantly to intuition (Ito 2008). Noticing the presence of complex patterns is one ability that requires intuition, and patients with focal cerebellar lesions often fail to

[8]This patient also had the expected deficit in tandem gait (see Chapter 9) and could not suppress her vestibulo-ocular reflex (VOR), which I found surprising given the location of the lesion. Possibly, her vestibular nuclei may have been affected by the tumor. In fact, several of the stroke patients in this study could have had Wallenberg syndrome (damage to the brainstem.

notice a pattern in a task that they are performing—one which normal subjects do notice (Molinari *et al.* 1997). The cerebellar hemispheres also seem to be involved in guessing the underlying structure of the Wisconsin card sorting test (WCST) (Cantelmi *et al.* 2008). The WCST is primarily used as a test of set-shifting, but in these subjects the deficit was probably in generating and testing hypotheses. Of course, both noticing a pattern and guessing a rule require working memory, another cerebellar function which could contribute to the deficits.

More specific tests of intuitive reasoning would be useful for evaluating the role of the cerebellum in this intellectual function. So far, the hypothesis that the cerebellum participates in intuition seems plausible. One can envision using the cerebellum to speed up cognitive processing during repeated exposure to the same problems, as in math classes throughout childhood. Mathematical problem solving can eventually become automatic, just like a skilled movement. Understanding of complex concepts (such as integration) can *become* intuitive, even though extensive linear reasoning was required to grasp the concept at its initial presentation. The cerebellum may play a part in these progressive changes.

## SPEECH AND LANGUAGE

Although some non-human primates have symbolic sounds, spoken language as such is thought to have appeared relatively late during hominid evolution, possibly coinciding with the dramatic neocerebellar expansion. Speech and language functions depend on Broca's area in the left frontal lobe and Wernicke's area in the left temporal lobe. Language is difficult to separate from other aspects of intelligence because language is needed to communicate intelligence, and intelligence is needed to use language. However, we can try.

### *Coordination for speaking*

Children sometimes stop speaking altogether for a time after surgery to remove cerebellar tumors. This phenomenon is known as "cerebellar mutism." Mutism seems to be due to damage to the posterior lobe vermis. Mutism recovers but is sometimes followed by severe stuttering, slow speech and/or a monotonous tone of speech, which persist (Riva & Giorgi 2000).

There is no question that the cerebellum is involved in the coordination of the muscles of the larynx, throat, tongue and lips to form the enormous variety of sounds that we need for speech. Speech sounds require a high degree of reproducibility, and such precision seems to be a likely job for the cerebellum. In fact, imaging studies generally indicate that the cerebellum is activated during the production of speech sounds (Cabeza & Nyberg 2000).

The anterior lobe also plays a part in speaking; it is active in generating speech, even when words are simply repeated or read aloud (Petersen & Fiez 1993). In fact, the anterior lobe is probably the part of the cerebellum that has the most significant role in articulating language (Timmann & Daum 2007).

Damage to the cerebellar hemispheres leads to another kind of motor speech deficit, both more lasting and more subtle than mutism. "Apraxia" is a neurological term that refers to a deficit of motor planning for the correct sequencing of movements. Some apraxias, such as gait apraxia and limb-kinetic apraxia, are routinely associated with cerebellar damage. Speech apraxia can also result from cerebellar damage. In his cerebellar gunshot patients, Holmes found that speech was highly abnormal (Holmes 1917). Patients with cerebellar cognitive affective syndrome also show abnormal quality of speech, such as high-pitched, "whining" speech (Schmahmann & Sherman 1998).

Cerebellar patients that do not have cerebral cortical damage can also have a particular kind of speech apraxia in which voice onset time is affected (Ackermann *et al.* 1999). The degree of impairment depends on the complexity of the task and the number of articulators (like the lips, tongue, and glottis) that need to be coordinated. Speech apraxia may be a problem with the timing of movements. The cerebellum may be necessary for accurate timing of the many movements involved in speech, as it is for the timing of other movements.

In addition to generating speech sounds, we use our cerebellar computers for the cognitive aspects of language and communication, such as word selection and grammar.

## *Word selection*

The anatomical pathways underlying language and the cerebellum are shown in Figure 11.1. Information related to the cognitive aspects of language may come to the cerebellum from the dorsolateral prefrontal cortex, by way of the pontine nuclei. Monkeys do not have Broca's area, and anatomical techniques are very limited in humans. Nevertheless, there is some evidence that Broca's area provides input to the cerebellum in humans. The anatomical arrangement is a closed loop; a projection from the right dentate nucleus reaches Broca's area in the left frontal lobe by way of the thalamus (Leiner *et al.* 1993).

There is evidence that, like Broca's area, the cerebellum is involved in non-motor as well as motor aspects of language. Many communication difficulties are experienced after cerebellar damage, and most of them can be divided into four groups: difficulties with motor coordination for generating speech; difficulties with organizing language (word finding, grammar, comprehension, and language processing); difficulties with working memory; and mutism.

The posterior lobe of the cerebellum is involved in the challenging task of finding the right word. When subjects are given nouns and asked to generate appropriate verbs for them, a region of the right hemisphere of the posterior lobe becomes more active (Petersen & Fiez 1993). Cerebellar cognitive affective syndrome often includes reduced verbal fluency. Word finding is more commonly affected than other language functions, and is affected in the majority of children who had cerebellar surgery to remove tumors (Levisohn *et al.* 2000). Some of the patients with cognitive affective syndrome are unable to generate particular words on demand (e.g., listing animals or names of cities) (Schmahmann & Sherman 1998).

Organizational and memory problems, that is, cognitive deficits, are most likely to result when the right hemisphere of the posterior lobe is damaged. One patient with damage to the right posterior lobe of the cerebellum had a particular problem with verbs; he could not generate them fluently, although his intelligence and language skills were good in general. He also could not improve with practice (Fiez *et al.* 1992). The restriction of some of these deficits to particular parts of speech indicates that the deficit is cognitive.

Several studies have implicated the cerebellar posterior lobes in verbal associative learning (Leiner *et al.* 1993) and in more general cognitive language functions, such as grammar. In cerebellar cognitive affective syndrome, nearly all of the patients that were tested showed language disability, including difficulties with grammar (Schmahmann & Sherman 1998). Other studies have reported agrammatic speech in patients with right cerebellar lesions (Silveri *et al.* 1998). In one study of children with cerebellar surgery, patients that had damage both to the vermis and to the right hemisphere had mutism followed by a persistent tendency to make grammatical errors; the problems with language organization were attributed to right hemisphere damage (Riva & Giorgi 2000).

Many of the patients in the surgical study were notably slow in performing all tests after cerebellar damage, consistent with the cerebellum's function of speeding up operations (Riva & Giorgi 2000). After cerebellar surgery in another study, children were sometimes slow both in speaking and understanding spoken language (Levisohn *et al.* 2000). In summary, word finding and language organization appear to rely, both for general execution and for speed, on the right hemisphere of the posterior lobe.

## *Working memory and rehearsal*

In a recent meta-analysis, a direct comparison of the cerebellar hemispheric regions that are activated during word selection and during verbal working memory showed that they overlap considerably but are not identical, with the word selection area centered slightly more lateral

(Stoodley & Schmahmann 2009). It appears that the hemispheric regions that are activated really do participate in non-motor aspects of language, because patients with restricted lesions of the posterior lobe hemispheres tend to have significant impairment of verbal working memory, vocabulary, and language comprehension (Tedesco *et al.* 2011).

Children who have had cerebellar surgery for tumors tend to show deficits in planning and in sequencing, as measured by tests of language skills. They often have expressive language deficits (Levisohn *et al.* 2000). After cerebellar surgery in another study, children were sometimes slow both in speaking and understanding spoken language (Levisohn *et al.* 2000). Following surgery on the right cerebellar hemisphere, auditory sequential memory and language processing are sometimes impaired. Both of these are cognitive functions that employ working memory (Riva & Giorgi 2000).

As already mentioned, some regions of the cerebellar posterior lobe may be part of a loop of reverberating activity that stores information phonologically ("rehearsal"). Obviously, rehearsal is useful when we have something very specific that needs to be said, or a listener with a short attention span. Its other uses include sentence comprehension and learning new words. The right posterior cerebellar hemisphere is activated during phonological rehearsal (Ackermann *et al.* 1999), along with Broca's area and other areas of the cerebral cortex.

The internal, hence very private, nature of phonological rehearsal has made it difficult to determine exactly what it is that the cerebellum contributes, but several interesting ideas have been proposed. One possibility is that the cerebellum organizes "inner speech," a brain representation of planned speech (Ackermann *et al.* 1999). During rehearsal, the cerebellum may compute the difference between actual and intended sounds, using an internal model; this would be consistent with its role in other kinds of movements, and might be considered motor planning for speech (Desmond *et al.* 1997).

As part of its role in inner speech, the cerebellum may also contribute to a "phonological output buffer," where words are stored immediately before being spoken. The output buffer is part of the rehearsal loop and also part of the speech-planning system (as it is usually envisioned). For example, one 18-year-old boy who had surgery to remove the right hemisphere of the posterior lobe had difficulty recalling and repeating lists of words. His deficit was believed to be in the phonological output buffer. The phonological short-term store, where the words are passively "held" for short periods, apparently was not affected (Silveri *et al.* 1998). All of this means that the job of the posterior lobe may be to get the words and the sentence structure that we are going to need next, ready.

The vermis may also contribute to cognitive language capabilities. A recent study of children who had undergone cerebellar surgery indicated that the vermis plays a broad role in language capability, including vocabulary, and also suggested a critical period, or vulnerable age, between 5 and 10 years. Events (such as surgery) occurring earlier or later than the critical period are believed to cause less trouble than damage during the critical period (Steinlin *et al.* 2003).

## POSSIBLE MECHANISMS

For motor systems, we have a wealth of information regarding how the cerebellum modulates and synchronizes its activity, makes use of sensory feedback, and times and corrects movements. But how does the cerebellum make its contribution to cognition? One possibility is that it contributes by means of *imagined* movements (e.g., during phonological rehearsal). But this idea can probably be rejected, because the cerebellar areas that are activated during inner speech (the posterior lobes) are not the same areas that are required for speech coordination (the anterior lobes). During imagined movements, we would expect all of the relevant motor areas to be active. The lack of overlap also suggests that during inner speech we are not imagining movement.

A better idea may be that the cerebellum adjusts the relative timing of different cognitive processes, in a way that is analogous to its role in coordinated movement. This sort of timing adjustment might be useful for correcting errors in cognitive processes (Strick *et al.* 2009). If so, this leads to the next question: if we use the cerebellum to improve cognitive performance, do we also take advantage of cerebellar plasticity to correct our cognitive mistakes? Conceivably, the cerebellum could correct many kinds of cognitive errors and could participate in implicit learning that improves cognitive processing.

How would it work? To address this question, we need to know how cognitive errors are encoded in the cerebellum. Of course, errors are generated by inappropriate actions, speech, or inadequate planning, but all of these are complex and may not produce stereotyped sensory inputs. Any feedback, whether it comes from the senses or an internal representation, will come much too late to depress the synapses that were involved in the original mistake. So how can cognitive errors be corrected? We are left with the possibility that a cognitive error signal must be used to correct cognitive errors; that is, the error signal is encoded somehow in abstract terms.

What form do the encoded errors take? One possibility is that the cerebellum contains forward models not only of motor systems but of cognitive systems, that is, the cerebellum learns to simulate the processing that is carried out by cortical areas (Ito 1993). In this hypothesis, the error would be the difference between the outputs of the cerebral cortical and cerebellar control systems. This means that the error would be generated at the same time that the behavior, speech, or thought is produced. The difference could be computed in the cerebral cortex (for example) and fed back to the cerebellum. The projection from the prefrontal cortex to the red nucleus, which in turn projects to the inferior olive (Figure 11.1) could carry the cognitive error signal. The simulation could then be corrected using similar cellular mechanisms to motor learning. Once the simulation has been fine-tuned, it can be used to generate the relevant behavior or thought with higher speed.

There are several potential problems with this hypothesis. For example, if the cerebellum is a faster processor, the outputs of the cerebral cortex and the cerebellum will not be synchronous unless something slows it down. Also, the cerebral cortex might continue to generate a (late) output and might be expected to interfere with the cerebellar output. But, the prediction that speed is the main contribution of the cerebellum could be tested.

## CONCLUSIONS

There is compelling evidence that the cerebellum has a major role in cognition, particularly in working memory and language. Changes in cerebellar activity occur during cognitive tasks, and lesions in the cerebellum can impair cognitive function. The posterior lobes of the human cerebellum contain regions that are uniquely human and that may be specialized for language and working memory. These regions are interconnected with the cerebral cortical areas that are involved in these two functions. Damage to the cerebellum causes major impairments in both working memory and communication, and cerebellar activity increases in situations that require them. The cerebellum does not perform either function on its own. Instead, it may serve to boost the processing speed and power of the cerebral cortex.

The cerebellum is relatively larger in modern humans than it was in early *Homo sapiens*, and its increased size may have allowed early humans to stay on top of their changing world. If so, we need it even more now. One hopes that we are on the verge of another expansion.

## REFERENCES

Ackermann, H., Graber, S., Hertrich, I. & Daum, I. (1999) Phonemic vowel length contrasts in cerebellar disorders. *Brain Lang.*, **67**, 95–109.

Allin, M., Matsumoto, H., Santhouse, A.M., Nosarti, C., AlAsady, M.H.S., Stewart, A.L., Rifkin, L. & Murray, R.M. (2001) Cognitive and motor function and the size of the cerebellum in adolescents born very pre-term. *Brain.*, **124**, 60–66.

Archibald, C.J., Wei, X., Scott, J.N., Wallace, C.J., Zhang, Y., Metz, L.M. & Mitchell, J.R. (2004) Posterior fossa lesion volume and slowed information processing in multiple sclerosis. *Brain.*, **127**, 1526–1534.

Cabeza, R. & Nyberg, L. (2000) Imaging cognition II: An empirical review of 275 PET and fMRI studies. *J. Cogn. Neurosci.*, **12**, 1–47.

Cantelmi, D., Schweizer, T.A. & Cusimano, M. (2008) Role of the cerebellum in the neurocognitive sequelae of treatment of tumours of the posterior fossa: An update. *Lancet Oncol.*, **9**, 569–576.

Carper, R.A. & Courchesne, E. (2000) Inverse correlation between frontal lobe and cerebellum sizes in children with autism. *Brain.*, **123**, 836–844.

Chen, S.H.A. & Desmond, J.E. (2005) Cerebrocerebellar networks during articulatory rehearsal and verbal working memory tasks. *NeuroImage.*, **24**, 332–338.

Dagher, A., Owen, A.M., Boecker, H. & Brooks, D.J. (1999) Mapping the network for planning: a correlational PET activaiton study with the Tower of London task. *Brain*, **122**, 1973–1987.

Desmond, J.E., Gabrieli, J.D.E., Wagner, A.D., Ginier, B.L. & Glover, G.H. (1997) Lobular patterns of cerebellar activation in verbal working-memory and finger-tapping tasks as revealed by functional MRI. *J. Neurosci.*, **17**, 9675–9685.

Dum, R.P. & Strick, P.L. (2003) An unfolded map of the cerebellar dentate nucleus and its projections to the cerebral cortex. *J. Neurophysiol.*, **89**, 634–649.

Exner, C., Weniger, G. & Irle, E. (2004) Cerebellar lesions in the PICA but not SCA territory impair cognition. *Neurology.*, **63**, 2132–2135.

Fiez, J.A., Petersen, S.E., Cheney, M.K. & Raichle, M.E. (1992) Impaired non-motor learning and error detection associated with cerebellar damage: A single case study. *Brain*, **115**, 155–178.

Glickstein, M. (2006) Thinking about the cerebellum. *Brain*, **129**, 288–292.

Habas, C., Kamdar, N., Nguyen, D., Prater, K., Beckmann, C.F., Menon, V. & Greicius, M.D. (2009) Distinct cerebellar contributions to intrinsic connectivity networks. *J. Neurosci.*, **29**, 8586–8594.

Hampson, D.R., Gholizadeh, S. & Pacey, L.K.K. (2012) Pathways to drug development for autism spectrum disorders. *Clin. Pharmacol. Therap.*, **91**, 189–200.

Holmes, G. (1917) The symptoms of acute cerebellar injuries due to gunshot injuries. *Brain*, **40**, 461–535.

Ito, M. (1993) Movement and thought: Identical control mechanisms by the cerebellum. *Trends. Neurosci.*, **16**, 448–450.

Ito, M. (2008) Control of mental activities by internal models in the cerebellum. *Nat. Rev. Neurosci.*, **9**, 304–313.

Joyal, C.C., Strazielle, C. & Lalonde, R. (2001) Effects of dentate nucleus lesions on spatial and postural sensorimotor learning in rats. *Behav. Brain Res.*, **122**, 131–137.

Kelly, R.M. & Strick, P.L. (2003) Cerebellar loops with motor cortex and prefrontal cortex of a nonhuman primate. *J. Neurosci.*, **23**, 8432–8444.

Kim, S.-G., Ugurbil, K. & Strick, P.L. (1994) Activation of a cerebellar output nucleus during cognitive processing. *Science.*, **265**, 949–952.

Lalonde, R., Joyal, C.C., Cote, C. & Botez, M.I. (1993) Simultaneous visual discrimination learning in lurcher mutatn mice. *Brain Res.*, **618**, 19–22.

Lalonde, R. & Strazielle, C. (2003) The effects of cerebellar damage on maze learning in animals. *Cerebellum.*, **2**, 300–309.

Leggio, M.G., Tedesco, A.M., Chiricozzi, F.R., Clausi, S., Orsini, A. & Molinari, M. (2008) Cognitive sequencing impairment in patients with focal or atrophic cerebellar damage. *Brain.*, **131**, 1332–1343.

Leiner, H.C., Leiner, A.L. & Dow, R.S. (1993) Cognitive and language functions of the human cerebellum. *T. I. N. S.*, **16**, 444–447.

Levisohn, L., Cronin-Golomb, A. & Schmahmann, J.D. (2000) Neuropsychological consequences of cerebellar tumour resection in children. *Brain.*, **123**, 1041–1050.

Lu, X., Miyachi, S. & Takada, M. (2012) Anatomical evidence for the involvement of medial cerebellar output from the interpositus nuclei in cognitive functions. *Proc. Natl. Acad. Sci. USA.*, **109**, 18980–18984.

Middleton, F.A. & Strick, P.L. (1994) Anatomical evidence for cerebellar and basal ganglia involvement in higher cognitive function. *Science*, **266**, 458–461.

Molinari, M., Leggio, M.G., Solida, A., Ciorra, R., Misciagna, S., Silveri, M.C. & Petrosini, L. (1997) Cerebellum and procedural learning: Evidence from focal cerebellar lesions. *Brain*, **120**, 1753–1762.

Mostofsky, S.H., Mazzocco, M.M.M., Aakalu, G., Warsofsky, I.S., Kenckla, M.B. & Reiss, A.L. (1998) Decreased cerebellar posterior vermis size in fragile X syndrome. *Neurology.*, **50**, 121–130.

Petersen, S.E. & Fiez, J. (1993) The processing of single words studied with positron emission tomography. *Ann. Rev. Neurosci.*, **16**, 509–530.

Rammani, N. (2006) The primate cortico-cerebellar system: anatomy and function. *Nat. Rev. Neurosci.*, **7**, 511–522.

Riva, D. & Giorgi, C. (2000) The cerebellum contributes to higher functions during development. *Brain.*, **123**, 1051–1061.

Schmahmann, J.D. & Sherman, J.C. (1998) The cerebellar cognitive affective syndrome. *Brain.*, **121**, 561–579.

Silveri, M.C., Di Betta, A.M., Filippini, V., Leggio, M.G. & Molinari, M. (1998) Verbal short-term store-rehearsal system and the cerebellum: Evidence from a patient with a right cerebellar lesion. *Brain.*, **121**, 2175–2187.

Steinlin, M., Imfeld, S., Zulauf, P., Bolshauser, E., Lovblad, K.-O., Luthy, A.R., Perrig, W. & Kaufmann, F. (2003) Neuropsychological long-term sequelae after posterior fossa tumour resection during childhood. *Brain.*, **126**, 1998–2008.

Stoodley, C.J. & Schmahmann, J.D. (2009) Functional topography in the human cerebellum: A meta-analysis of neuroimaging studies. *Neuroimage*, **44**, 489–501.

Strick, P.L., Dum, R.P. & Fiez, J.A. (2009) Cerebellum and nonmotor function. *Ann. Rev. Neurosci.*, **32**, 413–434.

Tavano, A., Grasso, R., Gagliardi, C., Triulzi, F., Bresolin, N., Fabbro, F. & Borgatti, R. (2007) Disorders of cognitive and affective development in cerebellar malformations. *Brain.*, **130**, 2646–2660.

Tedesco, A.M., Chiricozzi, F.R., Clausi, S., Lupo, M., Molinari, M. & Leggio, M.G. (2011) The cerebellar cognitive profile. *Brain.*, **134**, 3672–3686.
Timmann, D. & Daum, I. (2007) Cerebellar contributions to cognitive functions: A progress report after two decades of research. *Cerebellum.*, **6**, 159–162.
Weaver, S.H. (2005) Reciprocal evolution of the cerebellum and neocortex in fossil humans. *Proc. Natl. Acad. Sci. USA.*, **102**, 3576–3580.

## FURTHER READING

Cabeza, R. & Nyberg, L. (2000) Imaging cognition II: An empirical review of 275 PET and fMRI studies. *J. Cogn. Neurosci.*, **12**, 1–47.
Rammani, N. (2006) The primate cortico-cerebellar system: anatomy and function. *Nat. Rev. Neurosci.*, **7**, 511–522.
Stoodley, C.J. & Schmahmann, J.D. (2009) Functional topography in the human cerebellum: A meta-analysis of neuroimaging studies. *Neuroimage.*, **44**, 489–501.
Strick, P.L., Dum, R.P. & Fiez, J.A. (2009) Cerebellum and nonmotor function. *Ann. Rev. Neurosci.*, **32**, 413–434.

# 12 Sensing, Feeling, and Interacting

One of the major tasks of the brain is to assign importance to some of the sensory input that constantly bombards us, and ignore the rest of it. The avalanche of information includes everything we can discover about other people, as well as about other aspects of our environment. Meanwhile, the relevance of each sensory signal is linked to the emotions, if any, that it engenders. Emotions, along with our ability to regulate them, also contribute to how we behave in our interactions with other humans.

The cerebellum may be important in all of this. The connectivity of the human cerebellum with forebrain networks could enable it to participate in the processes of interpretation of sensation and feeling emotions, and also in determining how we will handle particular social situations. Of course, depending on their specific connectivity, different cerebellar regions participate in quite different functions. This means that the effects of cerebellar lesions that are caused by stroke or surgery, which have unique patterns of damage, will be variable. This fact causes great difficulty in replicating individual results. However, an increasing number of well-controlled studies, using positron emission tomography (PET) or functional magnetic resonance imaging (fMRI) in normal subjects, have yielded more consistent results.

As we have seen in the last few chapters, the cerebellar computer increases the speed and accuracy of a multitude of functions. In this chapter, I will round out the list of cerebellar functions, discussing experimental evidence that the cerebellum contributes to sensory perception and mental imagery; attention and prediction; music; social skills; mirroring; and, last but not least, emotion. Let's begin with two surprising observations about air hunger and thirst, which we may not think have anything to do with the cerebellum.

Thirst is a bodily sensation, but is also associated with emotion. Surprisingly, the anterior lobe vermis and hemispheres (specifically, lobule III) are activated during the subjective sensation of thirst. This activity is not correlated with actual dehydration. Lobule III is interconnected with

*The Cerebellum: Learning Movement, Language, and Social Skills*, First Edition. Dianne M. Broussard.

the cingulate cortex, which is also activated. Small, broadly-distributed regions of the hemispheres and vermis (lobules V–IX) also become activated during the development of thirst. It was proposed that the cerebellar activity reflects a vigilant state (Parsons *et al.* 2000). Of course, it could also reflect plans to go down the hall to the Coke machine. However, the interconnection with the cingulate cortex suggests a possible relationship between cerebellar activity and emotional state.

The same group also reported cerebellar activation during hunger for air. When carbon dioxide is inhaled, stimulating the urge to breathe, many regions of the cerebellum become activated concurrently with increased blood levels of carbon dioxide. Increased carbon dioxide activates several lobules of the vermis (I, II, IV, VII, and IX). Voluntary changes in respiration activate different areas. The results suggested that specific vermal regions are associated with the involuntary urge to breathe. Interestingly, a facemask produced a greater subjective sensation of air hunger than a mouthpiece in this study, and a comparison showed greater activation in the vermis of lobules III and IV with the mask. Activation of lobule III was also directly correlated with the subjective intensity of air hunger (Parsons *et al.* 2001).

What is going on here? In this final chapter, I will examine the evidence that the cerebellum participates in sensation, emotion, and self-control. All three could potentially participate in the cerebellar responses to these powerful physical urges.

## SENSORY PERCEPTION

The sense of touch is represented in the cerebellum. In rats, the dentate nucleus is dominated by a somatosensory representation of the face and whiskers (Parsons *et al.* 1997). In humans as well, activation of the dentate can depend on somatosensory discriminations that are made during a motor task. When the sensory stimulation is given alone, activation is still present. In fact, the sensory components cause more activation than the motor components do. Furthermore, the sensory activation is stronger when a sensory discrimination needs to be made (Gao *et al.* 1996).

Theories of the cerebellum tend to put the sensory and motor functions of the cerebellum together, by positing either that the sensory function of the cerebellum serves to guide movement or that its motor functions serve to get higher-quality sensory information. The role of the cerebellum in proprioception also depends on whether the sensory input occurs during active movement or during rest. For example, the cerebellum is always involved in estimating active force, but it may contribute even more when movement is ongoing (Bhanpuri *et al.* 2012).

Clearly, sensory information often guides movement. Joint-position signals are available during any limb movement, and we use them to guide

and correct the movement. The cerebellum could be involved in processing this information for perception of the hand's position in space. The processing power of the cerebellar circuitry might be advantageous, when we need to convert joint-position information into another coordinate system or reference frame. Consistent with the idea that we use the cerebellum for coordinate transformation, patients with lesions of the cerebellar hemispheres have difficulty identifying a shape that they trace while blindfolded (Shimansky *et al.* 1997). Alternatively, this observation might mean that the cerebellum is involved in visualizing the shape (mental imagery).

The cerebellum may be more involved in sensation than we realize. Sensory signals (proprioceptive, at least) are required for all motor tasks. What if, in some of the many motor tasks that have been studied, it is really sensory input that is causing cerebellar activation? The observation that sensory signals during a movement actually cause more activation in the dentate nuclei than the motor components do (Gao *et al.* 1996) is consistent with the idea that some regions of the cerebellum process mostly sensory information. In fact, it is possible that the posterior lobes, which receive large volumes of auditory and visual input via the pontine nuclei, may process sensory information for conscious perception. This would certainly fly in the face of current textbook descriptions of what the cerebellum does.

The timing functions of the cerebellum include, but are not limited to, motor tasks. Cerebellar patients who have deficits in the timing of movements (specifically, finger-tapping) also have deficits in their ability to perceive the duration of sounds (Ivry & Keele 1989). Trains of clicks with repetition rates above 4 per second activate a small zone in the right cerebellar hemisphere in a passive listening situation, suggesting a role for the cerebellum in auditory processing of short time intervals (Ackermann *et al.* 2001). Accurate perception of time intervals can improve with practice. Learned timing, based on any of the cerebellar mechanisms in Chapter 10, could be used for auditory perception as well.

The cerebellum does seem to participate in deciphering the complex temporal patterns of speech or music. Some patients with damage to the right cerebellar hemisphere cannot distinguish between words that differ by only a timing feature (such as "rabbit" and "rapid"). A small region in the right hemisphere (lobule VI) is activated in normal subjects during this task, and also during similar interval-discrimination tasks using noise stimuli (Ackermann *et al.* 2007). On the other hand, in cases of cerebellar mutism, which usually due to damage to the vermis, speech perception is generally normal. This means that auditory processing may be a function of the cerebellar hemispheres.

There is also enormous visual input to the cerebellum. One pathway originates from the superior colliculus and reaches lobules VI and VII of the vermis by way of the pons. Visual regions of the parietal and temporal cortex project, by way of the pons, to the cerebellar hemispheres, and these pathways carry information about visual motion. But the fact that visual information inundates the cerebellum does not necessarily mean that the cerebellum is involved in visual perception. Some of these visual signals probably guide movement directly in an automated fashion. For example, visual input to the ventral paraflocculus contributes to optokinetic gaze stabilization and pursuit eye movements. Visual inputs to the cerebellar hemispheres are thought to help in accurate reaching and stepping movements. Visual signals also guide motor learning in eye movement and other systems.

On the other hand, there is some evidence that the cerebellum also participates in perception of the speed and location of visual objects. Several studies have found deficits in visual motion perception in cerebellar patients. Patients with cerebellar lesions are less accurate than normal subjects in their judgment of the speed of moving objects (Leiner *et al.* 1993). In monkeys, lesions that included lobule VIII of the vermis caused impaired perception of visual motion in the absence of any motor impairment (Ignashchenkova *et al.* 2009).

The cerebellum receives abundant input from the parietal lobe, which serves a spatial "where" function, locating our bodies in space, as well as objects that we are interested in. Early reports suggested that patients with lesions in the left cerebellar hemisphere tend to have difficulty with cognitive operations in three-dimensional space (Leiner *et al.* 1993). Also, patients with congenital malformations of the posterior lobe hemispheres score below normal on cognitive tests that involve visuospatial abilities[1] (Tavano *et al.* 2007). A recent meta-analysis identified a region for spatial processing in the cerebellar hemisphere in lobule VI, on both sides (Stoodley & Schmahmann 2009).

The vestibulocerebellum is also intimately involved in the perception of self-motion in space. Its essential function in calculating the translational and tilt components of self-motion from vestibular sensory information was described in Chapter 9. The nodulus and uvula also receive visual information, which participates in our perception of self-motion and in navigation. Some patients with cerebellar cognitive affective syndrome experience vertigo triggered by unusual visual–vestibular conflict, and difficulty with tandem gait. These problems might be due to vermal or hemispheric damage to lobule IX.

Even the sense of smell may make use of the cerebellum. In studies of olfactory perception, the posterior lobe hemisphere is activated during

[1]More specifically, patients that could be evaluated using the Wechsler scale consistently did poorly on the object assembly subtest.

the presentation of odorants, even when no sniffing occurs. In contrast, the anterior lobe is activated during sniffing, whether or not an odorant is presented (Cabeza & Nyberg 2000). This strongly suggests an olfactory sensory function for the posterior lobe, and a separate (motor) function for the anterior lobe.

Of the many additional senses that humans do not possess, electroreception at least is known to make use of cerebellar-like structures in electric fish. Other examples almost certainly exist in nature.

So is the cerebellum sensory, or motor? By now, it should be obvious that this is the wrong question. The cerebellum has multiple functions, not a unitary function. Its functions include sensory perception and motor control. Its purpose is to contribute processing capacity to all of them.

## *Switching*

The cerebellar posterior lobe is involved in visuomotor learning. If it is working properly, subjects can learn to throw darts accurately while wearing prisms. Normally, the subject does not switch back to prism-free throwing without a period of trial and error. But it turns out that given enough time, monkeys can learn reaching both with and without prisms, and can do both in the same session. This suggests that the abstract concept "I am wearing prisms" enables us to switch control from one movement plan to another. Both the posterior-lobe cortex and the dentate nucleus are necessary for switching. Inactivation of *either* the dentate nucleus or large areas of the overlying cerebellar cortex (generally, this includes the lateral parts of lobules VII and VIII) prevents *both* performance of the learned task and new learning (Norris *et al.* 2011). Also, the cerebellar hemispheres are activated during the Wisconsin card sorting test (WCST), which requires directing attention to different features of the cards, based on abstract instructions (Cantelmi *et al.* 2008).

Switching between contexts is a crucially important cognitive function that we perform in daily life. One common example is getting in a car to drive. Once we are in the car, completely different motor plans are used to achieve the goals of motion through space, and perception of self-motion is also altered (for example, we can no longer rely on kinesthesia to tell us about motion through space). In a sense, the car requires its own frame of reference for both sensory and motor signals. Such contexts for learning are related to set-shifting, and are generally thought to be encoded in the forebrain. Possibly, that view should be re-evaluated.

Switching is a function that can involve a shift of attention, as it does in the WCST. But what about other aspects of attention? Is the cerebellum involved in directing attention to a given task or object in the first place?

## ATTENDING TO THE WORLD

Attention is one function of the executive network; it is necessary for carrying out almost any task. The very name "executive network" creates a bias to think in terms of doing something. However, the study of attention went on for several decades in animal models before the concept of the executive network became popular in neuroscience. During that time we thought of attention as being linked to sensory function. Either way, attention gets us ready for whatever is about to happen. It is a crucial factor in perception, learning, and cognition, as well as action. In a sense, attention controls a little window through which we experience the world in a given moment. To use it properly, we need to be able to direct our attentional window to one segment of our environment for reasonable periods. We also need to be able to shift the window to another segment when (for example) we detect a predator, or another vehicle when driving. Without this crucial pair of abilities, performance on any cognitive test will be poor, and we also would not pass the test of surviving in the natural world.

One cause of the social difficulties experienced by children with cerebellar cognitive affective syndrome is described as "inattention and irritability." The children themselves sometimes say that they "can't concentrate." Many meet the criteria for diagnosis of attention deficit hyperactivity disorder (Schmahmann *et al.* 2007), which is considered to be a problem of focusing attention.The question of whether the cerebellum participates in attention, or in related tasks, has been contentious. Unfortunately, attention can only be measured with respect to other tasks. Arithmetic, for example, is sometimes considered a test of attention, even though the task itself is calculation. As already noted, working memory and attention are especially difficult to distinguish.

### *Shifting and focusing attention*

The cerebellum may play a part in shifting attention to a new object or task. Children who have had cerebellar resection for tumors can show perseverative (repetitious) behavior (Levisohn *et al.* 2000). In a similar vein, children with cerebellar cognitive affective syndrome are sometimes diagnosed as obsessive–compulsive (Schmahmann *et al.* 2007). The adult with cognitive affective syndrome who had previously been a reader of classic literature, but had difficulty reading after her stroke, described her problem as being unable to hold more than a couple of concepts in her mind at a time (Schmahmann & Sherman 1998). This observation suggests that her problem could have been related to multitasking.

The cerebellar vermis is essential for accurate saccades, which direct our eyes toward new or interesting objects. However, just before making

the saccade, we first shift our attention to the new object. If attention is directed to a visual target, a saccade will usually (although not always[2]) be made to the target within about a tenth of a second. The shift of attention in space is, by definition, part of conscious awareness. Because they can be tested in well-established protocols, spatial shifts might be a good place to look for cerebellar involvement. Autistic children, posterior inferior cerebellar artery (PICA) stroke patients, and tumor patients with damage to the hemispheres are all slow to shift their attention to a different spatial focus. In the lesion cases, the important region for this deficit was the cerebellar hemisphere (Townsend *et al.* 1999).

In a study where no saccades were allowed, shifts of attention to different visual characteristics (form or color) of the same target activated regions of the posterior lobe hemispheres that were lateral to those involved in hand movements (Allen *et al.* 1997). Not only the cerebellar hemispheres, but also the vermis is involved in attentional shifts. Children with surgical resection of the cerebellum have problems with shifting attention between visual and auditory cues (Leiner *et al.* 1993). Patients with lesions that include lobule VIII vermis and hemispheres and lobule VI vermis are slow to shift attention covertly (i.e., without making saccades) to a different target location in space (Baier *et al.* 2010).

During attention shifts, attention is guided to novel or important features of the environment. This is the function of the salience detection network, which includes frontal and cingulate regions of the cerebral cortex. Discrete regions of crus I, crus II, and the dentate nucleus are functionally linked to the salience detection network in the resting state (Habas *et al.* 2009). The executive function network may also be important in attention. Exactly what the cerebellum contributes to these networks is not known, but by analogy with motor functions of the cerebellum, we can hypothesize that it does computations to regulate gain and adjust timing. This might help to ensure that in attention shifts are accurately directed at the correct feature or location, and that they occur rapidly and smoothly.

Picking "target" letters out of a stream requires selective attention and the ability to shift attention quickly. Patients with focal cerebellar lesions have problems when the two targets are presented close together in the stream, hence also in time (100–500 ms apart). This change does not seem to make the task more difficult for normal subjects. The surgical lesions in this study included part of one hemisphere (either right or left), primarily the lateralmost region of the posterior lobe (Schweizer *et al.* 2007).

There is reason to believe that the cerebellum is more important for shifting attention rapidly than it is for maintaining it. In one study, two

---

[2] An attention shift that is not accompanied by a saccade takes a little more effort and is called a covert attention shift.

tasks, shifting attention between shape and color, and sustained attention to either shape or color of the (foveal) target, were compared. The motor aspects of the tasks were the same keypress in both cases. Crus I showed greater activation during shifts of attention than it did during sustained attention. In this study, the activation related to attention shifts was bilateral, but greater on the left side of the posterior lobe (Le *et al.* 1998).

PICA stroke patients and autistic children have difficulty with attentional shifts. In addition, brain correlates of attention shifts are absent at short latencies after the cue to shift attention; this might reflect the loss of the cerebellar response to the cue. Given enough time, the patients do fine; they are simply slow to shift (Courchesne & Allen 1997). Of course, speed is crucially important for attentional shifts (think of driving).

Although shifting attention causes the largest activation, crus I and crus II may also participate in maintained attention to both the shape and color of the target. During maintained attention, activation in crus I and II is significantly reduced in autistic compared with normal subjects (Allen & Courchesne 2003). One possibility is that the deficits in shifting and maintaining attention may tend to occur together because some of the same cerebellar regions are involved in both functions. In fact, cerebellar input to the parietal cortex is modulated by focusing attention. In a visual discrimination task, there is more interaction between the cerebellum and cerebral cortex during maintained attention (Kellermann *et al.* 2012).

Working memory is active when focusing attention over time as well as during multitasking, when we keep two or more tasks in mind at one time. In multitasking, we may shift attention rapidly among cognitive tasks; whether we can actually process more than one task at a time is debatable. We already know that the cerebellum is involved in working memory, so it would be interesting to know whether it is also involved in the actual acts of focusing and shifting attention, apart from working memory. In fact, the cerebellar patients that showed a deficit in picking out letters did not necessarily show a significant deficit in a different test of working memory. This seems to indicate that attention is a separate function of the cerebellum (Schweizer *et al.* 2007).

## PREDICTION

For good execution of movement plans, the cerebellum predicts the likely outcome. It makes predictions about joint torques in order to fine-tune muscle contractions. Prediction of movements is also important for interpreting sensory events. For example, bumping your arm after moving it is quite different from having your arm touched by another person; the first is predicted, the second is not. In the sensation of "tickle," the cerebellum discriminates between sensations that are and are not due to our own movements. Only movements that an individual does not generate can contribute to his sensation of tickle (Fuentes & Bastian 2007).

The cerebellum responds selectively to various combinations of self-generated movement and sensory input. One reason may be that we need to use sensory feedback to adjust movements, and another may be that we use anticipated sensory feedback to adjust sensory perception. One study showed reduced activation in the right anterior cerebellar cortex during a self-generated movement that resulted in tactile sensory input, compared with a movement that did not result in any sensory input. The tactile input alone (generated by a robot) resulted in increased activation. In the same subjects, the somatosensory cortex was activated only when the tactile input was presented alone (Blakemore *et al.* 1998). These results support the idea that the cerebellum can determine whether a tactile input is self-generated or not, and modulate sensory perception accordingly.

## Gain control

If a sensory stimulus is too intense, we will not be able to process it fully, because the sensory circuitry will not be operating in the linear part of its range; we say it is "saturated." But if the sensory circuitry has advance information about how intense the stimulus will be, its response can be "dialed down" in time to stay within its linear range.

Although we cannot predict the future, the immediate past is quite useful in making guesses about the future. Sensory systems seem to evaluate the distribution of recent stimuli and adjust their gain accordingly. This process is known as gain control, or sensory adaptation. Unlike motor learning, sensory adaptation is immediate and does not persist. The cerebellar posterior lobe, which is linked to many sensory systems, is an obvious candidate to do the computation for sensory adaptation (Courchesne & Allen 1997). In support of this idea, stimulation of the posterior-lobe vermis can decrease both auditory brainstem responses and somatosensory-evoked potentials (Crispino & Bullock 1984).

By using gain control we can detect changes in sensory input, regardless of the background level of intensity. The cerebellum is involved in the representation of changes in somatosensory input as detected by event-related potentials. Patients with lesions that mostly included posterior lobe hemispheres were lacking the normal effect of novelty on the evoked potentials, which originate from the cerebral cortex (Restuccia *et al.* 2007).

A related sensory process is known as filtering. Filtering is what allows us to ignore noise and hear what our companions are saying in a crowded restaurant. For filtering, attention is important, but there may be other mechanisms that contribute as well. Pattern extraction, a process that early theories attributed to the cerebellum, may help us to listen in the right way. In autism spectrum disorders, the impairment of sensory filtering is sometimes given as an explanation for the fact that autistic individuals are hypersensitive to touch.

People who are accustomed to loud music in social situations may find it no longer interferes with conversation, but for many others, it definitely does interfere. This difference may be related to a sensory process called long-term habituation. Because of habituation, repeated stimuli, even loud noises like construction noises, that we are used to can be ignored much more easily. Habituation is well established with respect to vestibular stimuli; hence the phenomenon known as "sea legs."[3] It is also very pronounced in acoustic startle, which is caused only by an unexpected stimulus, not by a stimulus that occurs repeatedly. The caudal vermis is necessary for habituation or "suppression" of acoustic startle (Leaton & Supple 1986). Also, habituation of acoustic startle is abnormal in fragile X syndrome, as are other aspects of sensory gating (Frankland *et al.* 2004).

## MENTAL IMAGERY

We do not usually allow the cerebellum into the realm of the imagination; its function is supposed to be restricted to more mundane tasks. On the other hand, we have already suggested that the cerebellum has a role in narrative sequencing, which normally makes use of mental imagery. Mental imagery encompasses both sensory and motor functions. Different forebrain regions are activated during imagined sounds, images, and movements. However, the brain's default network is thought to be involved during all forms of mental imagery. The default network includes regions of the temporal and parietal lobes of the cerebral cortex, and also includes discrete regions of the caudal vermis and paravermis (lobule IX) (Habas *et al.* 2009). The cerebellar hemispheres are activated when subjects imagine finger-tapping or speech (Strick *et al.* 2009). Silent counting, a form of "inner speech," activates the cerebellar hemispheres and can be used to time events (Thach 2007).

In one study, subjects were asked to discriminate images of left and right hands that were presented very briefly in the central visual field. No movement was allowed during the task, and the motor cortex was not activated. But in the cerebellum, restricted regions of the posterior-lobe hemispheres were activated during the task. Meanwhile, in the forebrain, activation was seen in the posterior and inferior parietal cortex and the prefrontal cortex, in regions that are thought to be important for

[3]"Sea legs" refers to the ability of a sailor to walk on the swaying deck of a ship after a period of habituation. After returning to land, the sailor often has a little difficulty walking on a surface that is not moving.

perception of the body and space, and also for motor planning. The authors concluded that motor imagery, not sensory imagery, was used in the task. In other words, the subjects imagined a movement of their own hand to confirm a guess as to which hand was presented in the image (Parsons *et al.* 1995). This suggests that although mental imagery does not involve actual movement, it may sometimes involve internal models generated by the cerebellum.

## Music

A musician is not just pressing keys in sequence; she has a tune in mind. Making music involves hearing an imagined sound, and then making that sound using voice or instrument. If the tune is original, we call this process composition, and if it is memorized we call it performance; sometimes both happen at once. Mental imagery is required in each case. We have already seen that the lateral cerebellum is activated during generation of complex rhythmic sequences, and there is evidence that musical tempo and rhythm involves the cerebellum as part of a network (Levitin 2012).

So is the cerebellum involved in music, apart from the motor skills involved? Perhaps; the evidence is fragmentary but intriguing. The level of cerebellar activity during complex finger movements is reduced in musicians compared with nonmusicians. Classically trained keyboard players tend to have a relatively larger cerebellum than nonmusicians, and the relative size of the cerebellum is positively correlated with the intensity of practice (Hutchinson *et al.* 2003). A correlation does not identify cause and effect. Still, it is interesting to speculate that changes in size might be enabled when musical training begins in early childhood.

Playing music from memory activates different regions of the cerebellum than simply playing scales; scales activate mostly the anterior lobe (lobule V), but the music activates lobule VI hemispheres and deactivates the dentate nucleus (Parsons *et al.* 2005).

Reading music is a bit like language processing, and uses similar areas of the brain including the right cerebellar hemisphere. Musicians tend to have better verbal memory than nonmusicians. While listening to music, the expectation of a particular structure, which is important to our emotional experience of music, appears to involve the cerebellum (Levitin 2012).

Producing instrumental music requires working memory, attention, executive function, sequence learning, motor skill learning, eye–hand coordination, auditory discrimination, and auditory memory, not to mention the ability to precisely synchronize timing with other performers. Given this list, it seems likely that musical performance involves most of the cerebellum. But it is not clear whether the differences between musicians and nonmusicians are causally related to the music and, if so, which one is the cause and which, the effect in each case. At this point, we can tentatively conclude that musical performance, probably including the motor skills involved but also apart from them, is another function that requires the cerebellar computer.

## SOCIAL SKILLS

Social facility, or ability, can be considered a form of intelligence (Goleman 1995), which we can (at least conceptually) break down into four components—empathy, self-control, theory of mind, and communication. Empathy is sharing the emotions (or sensations) of another individual while self-control allows us to regulate our own emotions, or at least avoid expressing them when necessary. These two components are needed for smooth social interactions but are so intuitive that we do not consider them to be thought processes, and they are not included in most definitions of cognition. On the other hand, in order to have empathy, we do need to be able to identify the emotions that others are having. The ability to identify emotions is a cognitive skill, and could be included both in empathy and in theory of mind. Finally, theory of mind is an understanding of the intentions, motivations, and desires of other individuals.

Although the evidence that the cerebellum participates in the language skills that are required for communication is strong, theory of mind, empathy, and self-control are fundamentally different from language. Do they also require the cerebellum?

In autism, all four categories of social skills are defective; autistic children do sometimes display empathy, but they may have difficulty identifying emotions in others, partly, because of their problems with communication; for example, they may not be able to decipher nonverbal cues like the tone of voice. They also have difficulties with self-control.

A parallel situation occurs following mutations of the Tsc1 gene[4] in the Purkinje cells (P-cells) of mice. Like autistic humans, the mutants have reduced numbers of P-cells, and the P-cells they do have show abnormalities. In a test of social interaction, the mutants (unlike normal controls) show no preference for a novel mouse over a familiar one, and no preference for a novel mouse over a novel object. The mutants also spend considerably more time self-grooming, and infants emit more ultrasonic vocalizations, than wild-type mice. Cognitive inflexibility (another feature of autism) is demonstrated in these mice by their difficulty in learning a new platform location in the water maze, once one platform location had been learned (Tsai *et al.* 2012).

---

[4]Tsc1 is the gene for tuberous sclerosis. In humans, mutations of this gene are associated with autism spectrum disorders.

As we would expect if the cerebellum participates in social skills, patients with cerebellar cognitive affective syndrome often have social problems. Their difficulties can be classified into five groups: disorders of attention, of emotion, or of social skill set; autism; and psychosis. Social mistakes can include overfamiliarity, flamboyant and impulsive actions, and inappropriate comments. The behavior of these patients was described as regressive and childish (Schmahmann & Sherman 1998), like Phineas Gage after his accident.

### *Connecting with others*

One problem in cases of cerebellar damage may be related to difficulty with predicting how others will react, and/or with understanding their reactions. In recent years, there have been many imaging studies showing that the cerebellum responds to "emotional" images in a social context, such as faces, and also to recordings of voices with emotional intonation. A recent meta-analysis concluded that these social/emotional stimuli tend to activate small, discrete regions of lobules VI and VII and crus I (Stoodley & Schmahmann 2009). Other studies have concluded that both the cerebellar hemispheres and the vermis are involved in emotional processing for social situations.

One kind of stimulus that might be expected to activate empathic responses is an image of an injury. Such "empathic stimuli" tend to activate a circumscribed region in the hemisphere of lobule VI. Another region of VI responds to images that are thought to elicit drug craving (in addicts). A recent meta-analysis showed that these two regions are distinct and do not overlap, even though—and this is a key point—the images themselves are similar. The motor imagery, if any, associated with these images would also be expected to be similar. This suggests that the activation of lobule VI is associated with the actual emotion elicited, which is different in the two cases (Strick *et al.* 2009).

To interpret the movements of others, including facial expressions, we use the so-called "mirror neurons" in the forebrain. The regions of the cerebral cortex that contain mirror neurons in monkeys are interconnected with the dentate nucleus (Clower *et al.* 2001), suggesting that the dentate nucleus is involved in the interpretation of movement. Autistic individuals are impaired at interpreting biological motion, which may reflect a deficit in mirroring and may contribute to their social mistakes (Fuentes & Bastian 2007). The mirroring network is also useful for learning movement sequences by watching them, and this is another area where autistic individuals are impaired, consistent with the idea that mirroring is disrupted in autism (Fuentes & Bastian 2007).

In summary, evidence is accumulating that the cerebellum is involved in social skills. Although social skills may be considered a form of intelligence, they do not usually require linear thinking, but the more rapid kind of thought that we call intuition. The concept of the cerebellum as a pattern extractor has been used to argue that the cerebellum is responsible for intuitive thought (Ito 2008). Such intuitive thinking may be the basis for the role of the cerebellum in social skills.

## Emotional expression

Interpreting the emotions of others is necessary for regulating our own behavior. But we also need to be able to control our behavior. Based on its connections with the dorsolateral prefrontal cortex, we might expect that the posterior lobe is involved in regulating the expression of emotion during social interactions. When the cerebellum was believed to be purely motor in function, the possibility of a connection between the cerebellum and regulation of affect was ignored. In retrospect, though, the evidence has been around for a long time. After the posterior vermis was damaged by a scalpel cut along the midline in Botterell's 1937 study, the (formerly aggressive) monkey became fearful and ran away from experimenters (Botterell & Fulton 1938). Other studies found that cerebellar stimulation can evoke appetitive, social, and emotional behaviors, and can also modulate these behaviors. Cerebellar stimulation can lower the pain threshold, and can also enhance responses to noxious stimuli (Courchesne & Allen 1997).

Lack of self-control, and inappropriate expression of emotions, may be a major contributing factor in the poor social skills of cerebellar cognitive affective syndrome. The abnormalities of emotional expression are quite striking in a few cases of this syndrome. One patient actually hid under her bedsheets for a few days, following excision of a cerebellar midline tumor. Another patient, who had been an attorney before his stroke, began to swear frequently and attempted to kiss the examiner following a large PICA lesion. A third patient was described post-operatively as being "markedly blunt and inappropriately jocular." The families of all three patients reported persistent personality changes (Schmahmann & Sherman 1998). In cerebellar cognitive affective syndrome, lesions of the vermis can cause either "silliness," or "excessive giggling," or oscillations between apathy and crying. In fragile X syndrome, which also has a cerebellar component, chronic social anxiety impairs social interactions throughout life, and tantrums are common in childhood (Hampson *et al.* 2012).

Another syndrome known as "pathological laughter and crying" is associated with medial pontine lesions, which disrupt the cerebellar connections with nonmotor prefrontal areas. Patients with this syndrome laugh and cry inappropriately. Roughly one-third of patients with cerebellar atrophy also exhibit pathological laughter and crying (Schmahmann *et al.* 2007). Similar effects can follow cerebellar damage from surgery or stroke. One cerebellar patient had mild impairment of executive function, memory, and sustained attention, and a slight motor deficit that soon disappeared. He also showed pathological laughter and crying. An antidepressant stopped the attacks. The patient reported that he sometimes felt the emotion associated with his (inappropriate) behavior, not with the actual social situation (Parvizi *et al.* 2001). This seems to mean that emotion itself was disrupted.

Children with posterior fossa syndrome, which includes cerebellar mutism, also have inappropriate laughter and/or crying. These episodes are more prevalent when the lesion includes the posterior lobe vermis. In general, abnormal affect is typical after loss of all or part of the vermis, but not after damage to other cerebellar areas. These children remain markedly irritable and impulsive, according to their parents (Levisohn *et al.* 2000). Adults can also exhibit emotional lability after removal of tumors of the vermis.

## Emotional experience

Not only the expression of emotions, but also the actual experience of emotions, may be abnormal in cases of damage to the posterior-lobe vermis. Patients with congenital malformations of this region often have minor mood disorders, such as anxiety, and/or are socially withdrawn or overly aggressive. They are described as emotionally fragile. They also can show stereotyped, repetitive movements that are thought to indicate emotional distress (Tavano *et al.* 2007). Fragile X patients have heightened anxiety, and the vermis tends to be abnormally small (Mostofsky *et al.* 1998). In cognitive affective syndrome, the emotional aspects are more likely to occur when the vermis is included in the lesion. These problems may be classified either as disinhibition—the lawyer, for example—or as flattening of affect, that is, showing less emotion than one would expect. A few patients developed what appeared to be psychosis (Schmahmann *et al.* 2007). Together, these results indicate that the vermis is involved in regulating the emotional state.

Still more circumstantial evidence comes from premature infants, who often show reduced cerebellar volume throughout life. Later in life, these individuals tend to have a reduced sense of well-being, as evaluated by psychological testing, which correlates with the reduction in cerebellar volume (Allin *et al.* 2001). Emotional problems (anxiety and/or emotional fragility) can be associated with malformations not only of the vermis, but also of the cerebellar hemispheres (Tavano *et al.* 2007).

A recent study identified consistent cerebellar loci of activation as patients felt the emotions associated with recalling past events. Happiness, sadness, anger, and fear were evoked. Similar physiological changes (heart rate, etc.) were observed with all of the emotions. However, for each emotion, a different pattern of cerebellar activation was seen. Fear caused a large part of the vermis to be activated along with the right hemisphere; anger caused vermal and bilateral hemispheric activation; sadness caused right and left vermal and paravermal activation; and happiness caused slight activation of the vermis. These patterns were seen regardless of the particular memory involved. The results are especially significant because there was no motor task involved (Damasio *et al.* 2000).

In conclusion, now that many investigators have begun considering the possibility that the cerebellum is involved in sensation, emotion, and social skills, evidence is accumulating rapidly to support this possibility. In studies that are designed specifically to address such issues, it remains crucial to include controls for motor planning and other, more traditional cerebellar functions that might be involved in the behavior being measured.

## CONCLUSIONS

The posterior lobe of the cerebellum is a sensory as well as a motor structure. It is linked to the salience detection network, and assists with switching attention among objects or tasks. The posterior-lobe hemispheres are also useful in selecting a particular behavioral plan based on the overall context. Our senses must be adjusted for the level of intensity, and the cerebellum plays a part in such "gain control." Mental imagery appears to make use of the posterior lobes. We also employ them for some aspects of musical performance and perception. Even social skills, such as connecting with others and appropriate expression (or concealment) of emotions, require normal function of the posterior-lobe hemispheres and vermis. Evidence is accumulating that the vermis, in particular, is involved in actual emotional experience.

## REFERENCES

Ackermann, H., Riecker, A., Mathiak, K., Erb, M., Grodd, W. & Wildgruber, D. (2001) Rate-dependent activation of a prefrontal-insular-cerebellar network during passive listening to trains of click stimuli: an fMRI study. *Neuroreport*, **12**, 4087–4092.

Ackermann, H., Mathiak, K. & Riecker, A. (2007) The contribution of the cerebellum to speech production and speech perception: clinical and functional imaging data. *Cerebellum*, **6**, 202–213.

Allen, G. & Courchesne, E. (2003) Differential effects of developmental cerebellar abnormality on cognitive and motor functions in the cerebellum: an fMRI study of autism. *Am. J. Psychiat.*, **160**, 262–273.

Allen, G., Buxton, R.B., Wong, E.C. & Courchesne, E. (1997) Attentional activation of the cerebellum independent of motor involvement. *Science*, **275**, 1940–1943.

Allin, M., Matsumoto, H., Santhouse, A.M., Nosarti, C., AlAsady, M.H.S., Stewart, A.L., Rifkin, L. & Murray, R.M. (2001) Cognitive and motor function and the size of the cerebellum in adolescents born very pre-term. *Brain*, **124**, 60–66.

Baier, B., Dieterich, M., Stoeter, P., Birklein, F. & Muller, N.G. (2010) Anatomical correlate of impaired covert visual attentional processes in patients with cerebellar lesions. *J. Neurosci.*, **30**, 3770–3776.

Bhanpuri, N.H., Okamura, A.M. & Bastian, A.J. (2012) Active force perception depends on cerebellar function. *J. Neurophysiol.*, **107**, 1612–1620.

Blakemore, S.-J., Wolpert, D.M. & Frith, C.D. (1998) Central cancellation of self-produced tickle sensation. *Nat. Neurosci.*, **1**, 635–640.

Botterell, E.H. & Fulton, J.F. (1938) Functional localization in the cerebellum of primates. II. Lesions of midline structures (vermis) and deep nuclei. *J. Comp. Neurol.*, **69**, 47–62.

Cabeza, R. & Nyberg, L. (2000) Imaging cognition II: an empirical review of 275 PET and fMRI studies. *J. Cogn. Neurosci.*, **12**, 1–47.

Cantelmi, D., Schweizer, T.A. & Cusimano, M. (2008) Role of the cerebellum in the neurocognitive sequelae of treatment of tumours of the posterior fossa: an update. *Lancet Oncol.*, **9**, 569–576.

Clower, D.M., West, R.A., Lynch, J.C. & Strick, P.L. (2001) The inferior parietal lobule is the target of output from the superior colliculus, hippocampus, and cerebellum. *J. Neurosci.*, **21**, 6283–6291.

Courchesne, E. & Allen, G. (1997) Prediction and preparation, fundamental functions of the cerebellum. *Learn. Mem.*, **4**, 1–35.

Crispino, L. & Bullock, T.H. (1984) Cerebellum mediates modality-specific modulation of sensory responses of midbrain and forebrain in rat. *Proc. Natl. Acad. Sci. USA*, **81**, 2917–2920.

Damasio, A.R., Grabowski, T.J., Bechara, A., Damasio, H., Ponto, L.L.B., Parvizi, J. & Hichwa, R.D. (2000) Subcortical and cortical brain activity during the feeling of self-generated emotions. *Nat. Neurosci.*, **3**, 1049–1056.

Frankland, P.W., Wang, Y., Rosner, B., Shimizu, T., Balleine, B.W., Dykens, E.M., Ornitz, E.M. & Silva, A.J. (2004) Sensorimotor gating abnormalities in young males with fragile X syndrome and Fmr1-knockout mice. *Mol. Psychiat.*, **9**, 417–425.

Fuentes, C.T. & Bastian, A.J. (2007) 'Motor cognition'—what is it and is the cerebellum involved? *Cerebellum*, **6**, 232–236.

Gao, J.-H., Parsons, L.M., Bower, J.M., Xiong, J., Li, J. & Fox, P.T. (1996) Cerebellum implicated in sensory acquisition and discrimination rather than motor control. *Science*, **272**, 545–547.

Goleman, D. (1995) *Emotional Intelligence*. Bantam Books, Toronto, ON.

Habas, C., Kamdar, N., Nguyen, D., Prater, K., Beckmann, C.F., Menon, V. & Greicius, M.D. (2009) Distinct cerebellar contributions to intrinsic connectivity networks. *J. Neurosci.*, **29**, 8586–8594.

Hampson, D.R., Gholizadeh, S. & Pacey, L.K.K. (2012) Pathways to drug development for autism spectrum disorders. *Clin. Pharmacol. Therap.*, **91**, 189–200.

Hutchinson, S., Lee, L.H.-L., Gaab, N. & Schlaug, G. (2003) Cerebellar volume of musicians. *Cerebral Cort.*, **13**, 943–949.

Ignashchenkova, A., Dash, S., Dicke, P.W., Haarmeier, T., Glickstein, M. & Thier, P. (2009) Normal spatial attention but impaired saccades and visual motion perception after lesions of monkey cerebellum. *J. Neurophysiol.*, **102**, 3156–3168.

Ito, M. (2008) Control of mental activities by internal models in the cerebellum. *Nat. Rev. Neurosci.*, **9**, 304–313.
Ivry, R.B. & Keele, S.W. (1989) Timing functions of the cerebellum. *J. Cogn. Neurosci.*, **1**, 136–152.
Kellermann, T., Regenbogen, C., De Vos, M., Mossnang, C., Finkelmeyer, A. & Habel, U. (2012) Effective connectivity of the human cerebellum during visual attention. *J. Neurosci.*, **32**, 11453–11460.
Le, T.H., Pardo, J.V. & Hu, X. (1998) 4 T-fMRI study of nonspatial shifting of selective attention: cerebellar and parietal contributions. *J. Neurophysiol.*, **79**, 1535–1548.
Leaton, R.N. & Supple, W.F. (1986) Cerebellar vermis: essential for long-term habituation of the acoustic startle response. *Science*, **232**, 513–515.
Leiner, H.C., Leiner, A.L. & Dow, R.S. (1993) Cognitive and language functions of the human cerebellum. *Trends Neurosci.*, **16**, 444–447.
Levisohn, L., Cronin-Golomb, A. & Schmahmann, J.D. (2000) Neuropsychological consequences of cerebellar tumour resection in children. *Brain*, **123**, 1041–1050.
Levitin, D.J. (2012) What does it mean to be musical? *Neuron*, **73**, 633–637.
Mostofsky, S.H., Mazzocco, M.M.M., Aakalu, G., Warsofsky, I.S., Kenckla, M.B. & Reiss, A.L. (1998) Decreased cerebellar posterior vermis size in fragile X syndrome. *Neurology*, **50**, 121–130.
Norris, S.A., Hathaway, E.M., Taylor, J.A. & Thach, W.T. (2011) Cerebellar inactivation impairs memory of learned prism gaze-reach calibrations. *J. Neurophysiol.*, **105**, 2248–2259.
Parsons, L.M., Fox, P.T., Downs, J.H., Glass, T., Hirsch, T.B., Martin, C.C., Jerabek, P.A. & Lancaster, J.L. (1995) Use of implicit motor imagery for visual shape discrimination as revealed by PET. *Nature*, **375**, 54–58.
Parsons, L.M., Bower, J.M., Gao, J.H., Xiong, J., Li, J. & Fox, P.T. (1997) Lateral cerebellar hemispheres actively support sensory acquisition and discrimination rather than motor control. *Learn. Mem.*, **4**, 49–62.
Parsons, L.M., Denton, D., Egan, G., McKinley, M., Shade, R., Lancaster, J.L. & Fox, P.T. (2000) Neuroimaging evidence implicating cerebellum in support of sensory/cognitive processes associated with thirst. *Proc. Natl. Acad. Sci. USA*, **97**, 2332–2336.
Parsons, L.M., Egan, G., Liotti, M., Brannan, S., Denton, D., Shade, R., Robillard, R., Madden, L., Abplanalp, B. & Fox, P.T. (2001) Neuroimaging evidence implicating cerebellum in the experience of hypercapnia and hunger for air. *Proc. Natl. Acad. Sci. USA*, **98**, 2041–2046.
Parsons, L.M., Sergent, J., Hodges, D.A. & Fox, P.T. (2005) The brain basis of piano performance. *Neuropsychologia*, **43**, 199–215.
Parvizi, J., Anderson, S.W., Martin, C.O., Damasio, H. & Damasio, A.R. (2001) Pathological laughter and crying: a link to the cerebellum. *Brain*, **124**, 1708–1719.
Restuccia, D., Della Marca, G., Valeriani, M., Leggio, M.G. & Molinari, M. (2007) Cerebellar damage impairs detection of somatosensory input changes. A somatosensory mismatch-negativity study. *Brain*, **130**, 276–287.
Schmahmann, J.D. & Sherman, J.C. (1998) The cerebellar cognitive affective syndrome. *Brain*, **121**, 561–579.

Schmahmann, J.D., Weilburg, J.B. & Sherman, J.C. (2007) The neuropsychiatry of the cerebellum—insights from the clinic. *Cerebellum*, **6**, 254–267.

Schweizer, T.A., Alexander, M.P., Cusimano, M. & Stuss, D.T. (2007) Fast and efficient visuotemporal attention requires the cerebellum. *Neuropsychologia*, **45**, 3068–3074.

Shimansky, Y., Saling, M., Wunderlich, D.A., Bracha, V., Stelmach, G.E. & Bloedel, J.R. (1997) Impaired capacity of cerebellar patients to perceive and learn two-dimensional shapes based on kinesthetic cues. *Learn. Mem.*, **4**, 36–48.

Stoodley, C.J. & Schmahmann, J.D. (2009) Functional topography in the human cerebellum: a meta-analysis of neuroimaging studies. *Neuroimage*, **44**, 489–501.

Strick, P.L., Dum, R.P. & Fiez, J.A. (2009) Cerebellum and nonmotor function. *Ann. Rev. Neurosci.*, **32**, 413–434.

Tavano, A., Grasso, R., Gagliardi, C., Triulzi, F., Bresolin, N., Fabbro, F. & Borgatti, R. (2007) Disorders of cognitive and affective development in cerebellar malformations. *Brain*, **130**, 2646–2660.

Thach, W.T. (2007) On the mechanism of cerebellar contributions to cognition. *Cerebellum*, **6**, 163–167.

Townsend, J., Courchesne, E., Covington, J., Westerfield, M., Harris, N.S., Lyden, P., Lowry, T.P. & Press, G.A. (1999) Spatial attention deficits in patients with acquired or developmental cerebellar abnormality. *J. Neurosci.*, **19**, 5632–5643.

Tsai, P.T., Hull, C., Chu, Y., Greene-Colozzi, E., Sadowski, A.R., Leech, J.M., Steinberg, J., Crawley, J.N., Regehr, W.G. & Sahin, M. (2012) Autistic-like behaviour and cerebellar dysfunction in Purkinje cell *Tsc1* mutant mice. *Nature*, **488**, 647–651.

## FURTHER READING

Courchesne, E. & Allen, G. (1997) Prediction and preparation, fundamental functions of the cerebellum. *Learn. Mem.*, **4**, 1–35.

Fuentes, C.T. & Bastian, A.J. (2007) 'Motor cognition'–what is it and is the cerebellum involved? *Cerebellum*, **6**, 232–236.

Damasio, A.R., Grabowski, T.J., Behara, A., Damasio, H., Ponto, L.L., Parvizi, J. & Hichwa, R.D. (2000) Subcortical and cortical brain activity during the feeling of self-generated emotions. *Nat. Neurosci.*, **3**, 1049–1056.

Strick, P.L., Dum, R.P. & Fiez, J.A. (2009) Cerebellum and nonmotor function. *Ann. Rev. Neurosci.*, **32**, 413–434.

# What Does the Cerebellum Do?

The cerebellum is a powerful processing device that we can use to improve the performance of our brains in a multitude of tasks. The cerebellum has broad connectivity and powerful intrinsic computational mechanisms that have the potential to help the forebrain with motor, sensory, cognitive, and social demands. The components of the cerebellum are arranged in a loop that is continuously active, ready to help out. Different cerebellar modules make recurrent connections with the brainstem and with the forebrain, including the parietal and frontal lobes.

The cerebellum does computations on a scale that is found nowhere else in the brain. For this purpose, it has enormous numbers of neurons and synapses, as well as some of the largest and some of the smallest neurons in the brain. It has several specialized structures: layers of parallel axons; other axons that climb like ivy; dendritic arbors that resemble sea fans; glomeruli; and giant synapses. Even its ion channels are unusual.

Theories of cerebellar function make use of the efficient, highly stereotyped, modular structure of the cerebellum for learning, for pattern recognition, and for timing. Modelers have tested several of these theories for plausibility, and some are also supported by experimental evidence. Key features of the successful models are synaptic plasticity; synchrony of firing; and filtering.

Synaptic mechanisms for learning and memory in the cerebellum include cerebellar LTD and several other kinds of plasticity in the cerebellar cortex and in the deep cerebellar nuclei. Most, if not all, of the synapses in the cerebellum are plastic and, at least, some of the neurons can also modify their intrinsic excitability. Using such mechanisms, the cerebellum learns to carry out new motor tasks and to correct existing ones, making them quick and accurate. Inhibitory interneurons in the cerebellar cortex, including the unique basket cells, play a major role in timing cerebellar responses and in the learning process itself. Timing also depends on cells in the inferior olive that are interconnected in synchronous groups.

In collaboration with other brain systems, the cerebellum uses sensory signals and internal models to control movement very precisely, time it accurately, and correct it immediately, contributing its computational power to make our movements efficient and accurate. Coordinated movement is a very complex endeavor, and is absolutely necessary for survival. The

*The Cerebellum: Learning Movement, Language, and Social Skills*, First Edition. Dianne M. Broussard.

cerebellum solves some of the more difficult aspects of movement, such as coordination across joints and directing movement at targets in the outside world. It balances us on our two (rather small) feet and allows us to leap, twirl, and run down stairs if we feel like it. Our manual dexterity may be largely due to the cerebellum and its collaboration with the cerebral cortex. Best of all, during routine tasks such as walking, it leaves our thoughts completely free to concentrate on other issues.

But the cerebellum does not stay out of our thought processes. In fact, the increase in speed and accuracy that the cerebellum provides may extend across all realms of brain function. The posterior lobes and ventral dentate nucleus, which are hugely expanded in humans, seem to have taken on many non-motor tasks that were previously thought to be allocated exclusively to the cerebral cortex. This is a busy area of research, and in Section IV, I have outlined some of the exciting ideas that are beginning to garner support. The cerebellum may be involved in processes as diverse as emotion, language, working memory, and social behavior. I expect that in the next few years, considerably more will be learned about the functions of the human cerebellum.

I was in graduate school in 1979, and learned about the cerebellum from Maurizio Mirolli, a talented and inspiring teacher. But I found it difficult to grasp just what it is that the cerebellum does. Before his next lecture, inundated by questions in the hallway (I was not the only one!), Dr. Mirolli emphatically reiterated his point: "The cerebellum does EVERYTHING!" This could not possibly be true, I thought. But in fact Mirolli, like Masao Ito and others, made an intuitive leap that has, largely, been supported. As I hope this book has convinced you, most brain functions, if defined broadly enough, make use of the cerebellum.

# Index



*The Cerebellum: Learning Movement, Language, and Social Skills*, First Edition. Dianne M. Broussard.

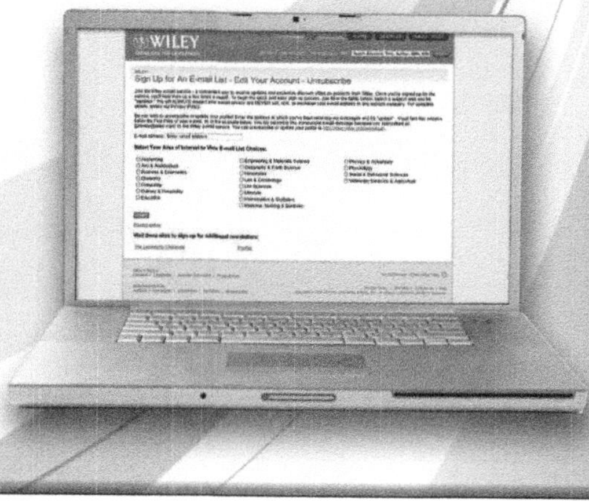
WILEY

Printed and bound by CPI Group (UK) Ltd, Croydon, CR0 4YY
17/09/2026
14964289-0003